THE ELECTRONIC PROJECT BUILDER'S REFERENCE

DESIGNING AND MODIFYING CIRCUITS

For all my friends at Gernsback,
and for Wayne Green as well.

THE ELECTRONIC PROJECT BUILDER'S REFERENCE

DESIGNING AND MODIFYING CIRCUITS

JOSEF BERNARD

FIRST EDITION
FIRST PRINTING

Library of Congress Cataloging in Publication Data

Bernard, Josef, 1943-
The electronic project builder's reference : designing and modifyig circuits / by Josef Bernard.
p. cm.
ISBN 0-8306-9260-6 ISBN (invalid) 0-8306-3260-X
(pbk.)
1. Electronic apparatus and appliances—Amateurs' manuals.
2. Integrated circuits—Amateurs' manuals. I. Title.
TK9965.B47 1990
621.381—dc20 89-20472
CIP

TAB BOOKS Inc. offers software for sale. For information and a catalog, please contact TAB Software Department, Blue Ridge Summit, PA 17294-0850.

Questions regarding the content of this book should be addressed to:

Reader Inquiry Branch
TAB BOOKS Inc.
Blue Ridge Summit, PA 17294-0214

Acquisitions Editor: Roland S. Phelps
Production: Katherine Brown

Paperback cover photograph: Josef Bernard

All photographs by Josef Bernard unless noted otherwise.

Contents

ACKNOWLEDGEMENT

A note of thanks to my friend TJ Byers, who helped clear up a number of points for me. If I misinterpreted him, I'm to blame.

Introduction

The world is full of eternal mysteries: the Sphinx, time, gravity, and—to large numbers of people—electronics. It doesn't matter whether you're just getting started or whether you've been pushing electrons for years. There are probably some things that would be a lot easier for you to understand if you just had a few simple answers. Alas, no one seems to have them. Or, if he does, he's not talking.

Why, for example, are there so many different kinds of transistors? How do you choose the value of a current-limiting resistor? And, for that matter, why do you *need* a current-limiting resistor? And why, in the equation representing Ohm's law, are volts represented by E and amperes by I? Why not V and A?

Somewhere there are reasons for all of these things. If you know them, both the art and the science of electronics become much more manageable.

There are plenty of books on the shelves that tell you how to build electronic gadgets and gizmos. They give you a list of parts, maybe even suggest a place to buy them all at once, and then tell you how to put them all together. "Insert tab *A* into slot *B*," they say, "then put peg *C* into hole *D*." The words "tab" and "slot" might be replaced by "MOSFET" and "heat sink" or by "integrated circuit" and "socket," but in the end it's the same.

What you get is a set of instructions for putting something together. What you don't get is an explanation of how or why it—whatever it might be—works, or what to do if it doesn't work. If you're lucky and there are no mistakes in the instructions and you make no mistakes yourself in putting the device together, your whatever-it-is works. Nevertheless, you might not have the faintest idea why it works.

There are also plenty of books that tell you how things work. Or, to be more precise, they teach you the principles on which things work—the natural laws and the formulas and other things you need to pass tests—but they don't tell you how to apply that information to the real world. There are lots of people who can work out calculations in their heads involving Ohm's law, but some of them would probably be hard pressed to tell you what they were going to do with the result.

Although this sort of problem is probably common to many fields, electronics books in particular are frustrating in this respect. They show you either the "how" or the "why" or the practical or the theoretical, but rarely do they put the two together to show a cause-and-effect relationship. This can be particularly frustrating for the newcomer to electronics who is deluged with theory and wants to relate it to the real world around him. Theories rarely tell you how to use information, and the books that show you how to build things rarely tell you why you are doing what you're doing or how to do it a little bit differently so as to achieve the precise results you want—or just to satisfy your curiosity.

This book, I hope, will be different.

This is not a project book. If you're looking for cut-and-dried plans for a 50-LED pendulum clock or a remote-control device for your clock radio, you're in the wrong place. There are some circuit diagrams in this book, and the circuits they illustrate work, but the circuits are present more to illustrate the concepts presented than for your amusement. They also give you a starting point for designing your own circuits using the principles presented.

This is by no means a textbook on electronics, either. It's a colloquial ramble—almost a series of monologues, if you will—through the province of electronic design and construction. It is not an all-in-one book for beginners. You will undoubtedly find it worthwhile to refer to other, "drier" books for some of the material you will want to know. However, this low-key little work might surprise you and provide you with answers you'll be hard pressed to find elsewhere. When you need information about a point that arises, check the index here and you may very well find what you're looking for.

You might not find a lot of things to build here, but you will find a lot of advice. Isn't that what you bought this book for? Use it well!

Josef Bernard

PART ONE

1
Tools of the Trade

Getting a start in electronics need not be an expensive proposition. One of the electronics experimenter's most valuable possessions in his junk box. And one of the first things he learns is: NEVER THROW ANYTHING OUT! The junk box is where all the leftover components, the bargain components, and the recycled components wind up. The same goes for tools. Eventually you'll find yourself with extras and duplicates, perhaps tools that are still in fine condition but that you've replaced with higher-quality ones or ones that do something a little better than the old ones did. Don't get rid of the old ones; you'll find a use for them someday.

For whatever reason, junk boxes seem to grow until one day you're suddenly forced to wonder where all that stuff came from. We'll talk more about junk boxes and how to start building your own later. For the moment, though, we have to assume that you haven't anything on hand but want to get started. What are you going to need and where are you going to get it?

THE BARE NECESSITIES

If you're just getting started in electronics and have no other resources at the moment, you don't need anything but this book in hand. If you've already tried learning electronic theory from other books but gotten frustrated because they weren't clear enough or were too complicated, what you're reading now will help. There are a lot of subjects this book does not touch upon, but you'll also find a lot that can help you to understand what you find elsewhere.

It is always useful not to rely on just one source, no matter how authoritative it may claim to be, but to hear the same story from several different sources told in several different versions. One of those versions will say something to you the others don't. All of a sudden the pieces will fall into place and become clear. The problem is that you never know which source will be the one to do that.

Furthermore, this type of revelation is seldom the payoff from a single source. Each piece of information you obtain tends to reinforce all the others to the point that it is sometimes difficult to say just what it was that caused you to "see the light" in the first place. This book can be just one of those many sources. But who knows, maybe it will be the one to clarify the others for you.

More Necessities

What got you interested in electronics in the first place? It was probably someone who was already an electronics hobbyist (maybe a ham, for example), or someone employed in the electronics industry. Or maybe it was just someone, like yourself, who wanted to know how things worked and got hooked while looking for the answers.

Whatever (or whoever) was responsible, if he plays around with electrons at all, he no doubt has a bigger junk box than you have at this stage of the game. He can probably provide you with what you need to get started at no cost to either of you, and will, indeed, probably be glad to help you begin.

While you could leaf through this book and make a list of all the components used in all the projects, and then order them by mail or go out to a nearby parts store (if you're fortunate enough to have one nearby) and ask them to fill it, you would be duplicating a lot of effort and wasting some money if you did. The idea here is to build simple circuits that demonstrate the principles of electronic design, and then to modify the circuits a bit to perform slightly differently to see how these principles can be applied in practice. It would make no sense to build each circuit to last forever because you're going to play around with it and change this or that—probably again and again. Therefore, the parts you have can and should be reused. If you were to go out and buy all the 4.7k resistors called for individually in this book, you'd wind up with quite a few too many. These could, of course, go into your junk box for use at some later time, but it would make more sense to start with a smaller inventory and recycle it.

SOMETHING TO BUILD ON

Just about any piece of electronic equipment you look into today is built using printed circuit techniques. The components that make up the device are mounted on a sheet of insulating material and connected to one another by thin traces of copper foil that carry electrical power and signals among them. Printed circuit boards, known simply as PC boards, are great when it comes to compactness, convenience, and ease of assembly. They lend themselves very nicely to automated assembly techniques where everything—including the insertion of the components into the boards and the soldering of their leads to the boards—is done by machine. Even for hand assembly, PC boards are hard to beat.

However, PC boards have to be designed and manufactured before you can start stuffing them with components. While this is not an impossible procedure, it generally means too much work if you're just building one-of-a-kind circuits. PC boards are best suited to volume production. Once you've done the hard part of designing the circuit and laying it out on the board, what follows becomes correspondingly simpler. So, while you may want to build the final version of your design using PC construction techniques, you will undoubtedly want to do your initial prototyping using more flexible methods better suited to one-of-a-kind projects.

The layout and construction of circuit prototypes is frequently called *breadboarding*. This method got its name in the earlier days of electronics experimenting when facilities were even more limited than yours seem to you now. Many a circuit was literally built on a breadboard, which served as a combination PC board of sorts and chassis. And, though the materials have changed, the term remains.

Today's breadboards are considerably more convenient to use. One type in common use consists of a plastic plate with a gridwork of holes into which wires and component leads can be placed; small springlike connectors grip them snugly when they are inserted. Beneath the plastic runs a network of flat metal connectors that can be used to carry supply voltages or signals. Generally, there are one or more power and ground buses. A solderless (because no soldering is required to make connections) breadboard of this type is illustrated in Fig. 1-1.

Most breadboards of this sort are especially intended to accept DIP (dual in-line package) integrated circuits and have a center channel that these packages straddle when in place. The holes for component leads are spaced at 0.1-inch intervals. This tenth-of-an-inch spacing conforms to that of the leads on DIPs and on other components used with them.

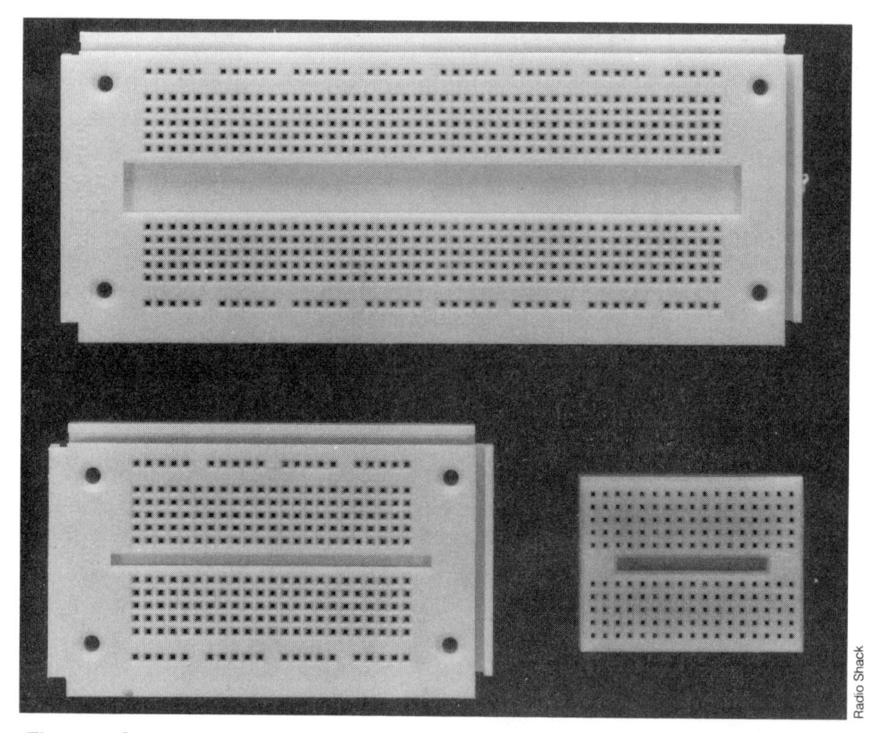

Radio Shack

Fig. 1-1. Solderless breadboards are a great convenience in prototyping circuits. The wide trough in the one at the top of the photograph is to accommodate ICs with 0.6-in spacing.

Power and signal buses run at right angles to the direction of the IC package, allowing you to connect other components or wires to the pins of the IC. There are two sets of buses (one on either side of the center channel). The power buses are usually located at the edge of the breadboard, and are perpendicular to the others.

Solderless breadboards come in several different sizes and degrees of complexity. The smallest are big enough to hold just a single, small IC and perhaps a couple of other components. Larger ones will accept 24-pin or larger packages with room to spare. Some of these breadboards come with screw terminals for connecting the power buses easily to a power supply; others require you to use one of the small holes providing access to the power bus to connect a wire to the power supply.

Some really elaborate solderless breadboards come mounted on a chassis that contains a power supply and perhaps several potentiometers or other devices ready to be connected to your circuit (Fig. 1-2). While an apparatus like this is nice to dream about, and even nicer to have, it is

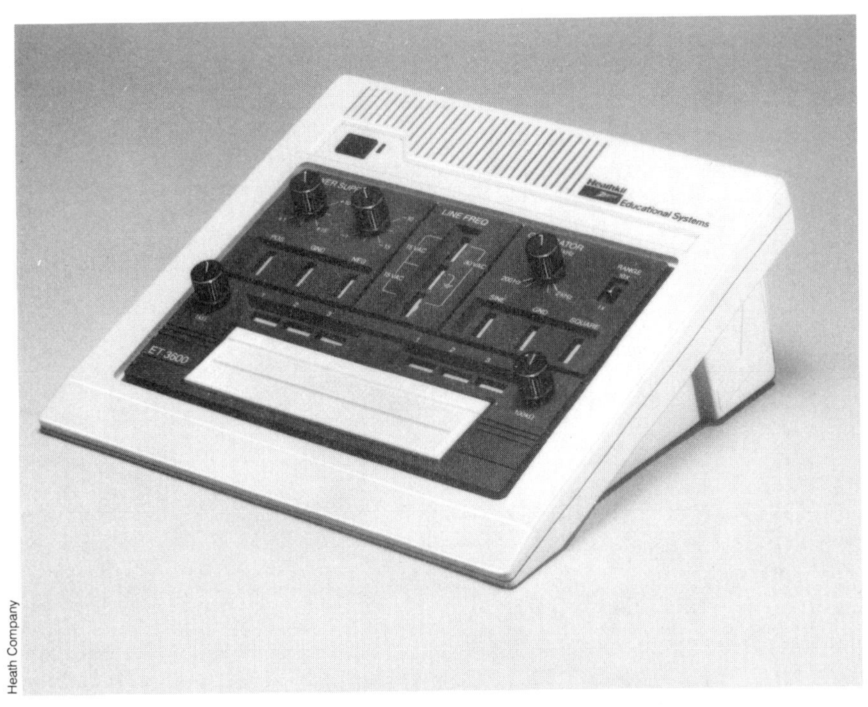

Heath Company

Fig. 1-2. Some electronics "trainers" include a solderless breadboard, power supply, potentiometers, and other electronic circuit elements.

frequently more practical to use something less complicated. The values the manufacturer chose for the built-in components might not be the ones you need for your purposes, and the money you save by not buying what you are not going to use can be well spent on other things.

For the purposes of this book, a plain, medium-size solderless breadboard will do. It will allow you to build the circuits we discuss, and then tear them down again so you can reuse the components later. That's what's nice about the "solderless" part—as long as you don't cut the component leads too short you can just pull the parts out and use them over again.

Of course, solderless breadboards are not the only means of construction open to you. Other techniques that are nearly as simple to use are also more permanent because they require the use of solder. There is no reason why you can't use them if you don't mind rendering some of the components not fit for reuse.

Perforated construction board has been an experimenter's standby for a number of years. It is a sheet of insulating phenolic material, usually $^1/_{16}$ of an inch thick, with a regularly spaced grid of holes drilled through it (Fig. 1-3). Most of the perfboard you'll find uses the same hole spacing—0.1-inch—as the solderless breadboards mentioned earlier. There is also a type with slightly larger holes at 0.265-inch intervals

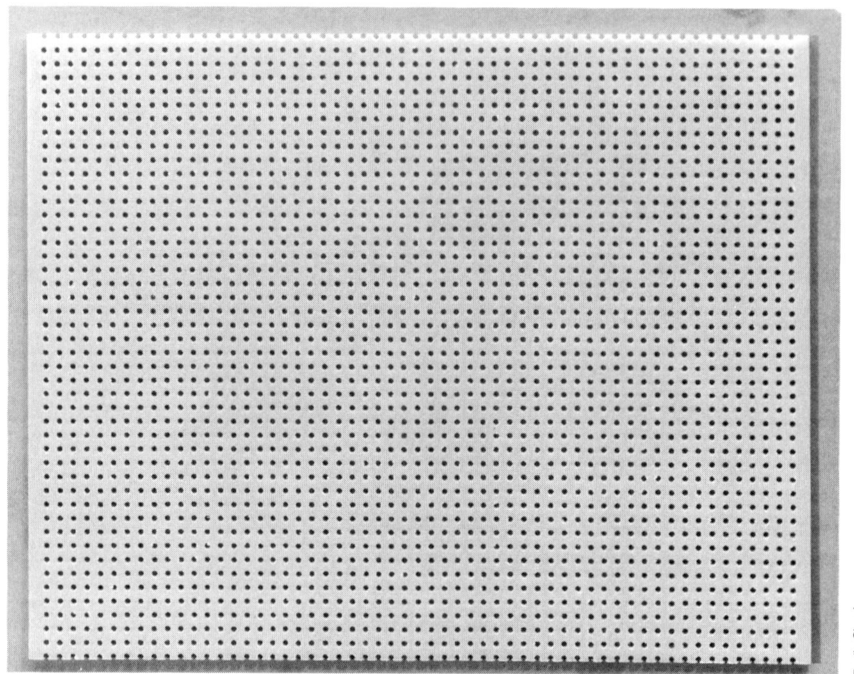

Radio Shack

Fig. 1-3. Perforated construction board can be the next (and sometimes final) step in design after a circuit has progressed beyond the breadboarding stage. The material can easily be cut and drilled for permanent mounting in an enclosure.

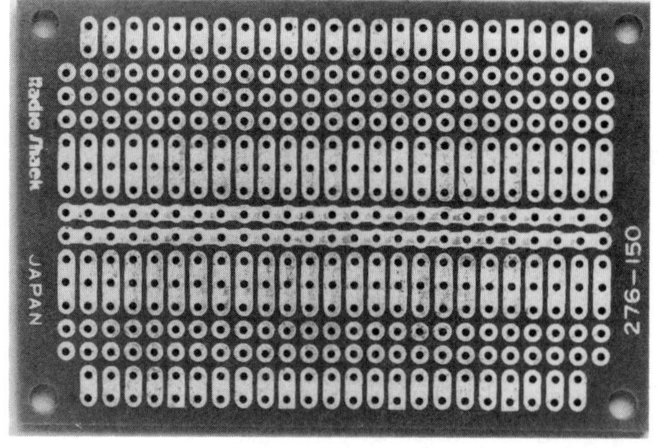

This kind of construction board, with pads on the underside to which components can be soldered, is more convenient to use than plain perforated construction board. However, the layout of the solder pads may restrict the layout of your components.

that you should try to avoid because not many components made these days use that spacing for their leads or pins.

Although a few components—some miniature potentiometers for instance—use the 0.265-inch spacing, they are rare. Watch out for them when you're out shopping. If you have to, you can usually use them on 0.1-inch-spacing perfboard. Some work may be required to make them fit.

If you buy a large sheet of perfboard you can use a saw to cut it into smaller pieces suitable to the size of each of your projects. You can also use perfboard to build the final versions of your circuits after you have them working just the way you want.

The problem with perforated construction board is that is is not really reusable and, because construction on it involves soldering, the components mounted on it are difficult to salvage and reuse. More about that later, though.

There are also a number of types of perforated construction board with copper pads on one side. These are a little easier to use than the plain variety. Component leads can be soldered to the pads rather than being twisted together and soldered to one another. This type of board is also suitable only for one-time use and is really little more than an expensive luxury.

If you plan to do a lot of experimenting and playing around with your prototypes, make the initial investment in a reusable solderless breadboard. In the end it will save you quite a bit of money in terms of recycled components, and will give you the flexibility you want if you're going to take full advantage of this book.

If, on the other hand, you are going to build the projects here exactly as shown, once and for all, without any intention of ever changing any of them to suit your whims or satisfy your curiosity, then you may as well use a more permanent means of construction.

FINDING PARTS

If you're fortunate, you have one of those friends with the well-stocked junk box mentioned earlier. Some evening or weekend afternoon you can go over to his place with your shopping list in hand and browse through and pick up what you need or whatever catches your eye. Not everyone is this lucky, though. And not everyone—hardly anyone, in fact—has a neighbor with a collection of surplus neatly arranged for browsing so you can pick or choose.

As you'll discover when you've acquired your own junk box, such collections are not known for their orderliness. Mostly they consist of cardboard cartons filled with . . . well, stuff. At the top are probably old circuit boards, built and discarded, the components stuffing them waiting to be recycled. Lower down are jars or plastic containers with knobs, switches, and the like amassed here and there and waiting to be used. Way at the bottom is the heavy stuff, transformers for instance, and maybe a few heat sinks. And scattered throughout, like pebbles in concrete mix, are loose resistors and capacitors, slowly filtering their way down to the bottom like silt in a lake.

After a while you'll probably get to know your junk box intimately. As you're searching for a particular component, you'll notice yourself saying something like, "I saw it in here just last summer . . . it was right underneath that big electrolytic capacitor I was saving for the power supply and wound up giving away to . . . *aha!*—here's that potentiomenter I wanted last month; I *knew* I shouldn't have bought a new one . . . ." A trip through your junk box can be a trip down memory lane, bringing to mind recollections of places, projects, and friends.

Buying Components

A great place to find parts cheap is at ham flea markets. Sometimes, too, parts vendors turn up at computer shows.

Flea markets are great sources of power-supply components—transformers, big capacitors, heat sinks, and the like. The transformers aren't always marked but you can get a good deal if you know what you want and keep your eyes open for something that looks like it might fill the bill.

Some computer shows—the kind known as "selling shows" that are open to the public—are a sort of flea market consisting mainly of booths where people sell such things as IBM-computer clones, plug-in boards, and peripherals such as disk drives and printers. These shows frequently attract vendors of components, who make the weekend show circuit a part of their business. The vendors, who are also frequently found at ham flea markets, usually have their stock well organized, and are a particularly good source of brand-new semiconductors from diodes and LEDs to microprocessors.

Another thing you might come across at flea markets are surplus circuit boards. These have no practical use in and of themselves, but can serve as a quality source of ICs, potentiometers and medium size capacitors. If you're looking for ICs to reclaim, make sure they're in sockets. Integrated circuits that are soldered directly to the board are difficult to unsolder. They can be destroyed by the heat required to get them free.

You can sometimes also reclaim transistors and other components from these boards, but doing so isn't really worth the trouble. Aside from the fact that the salvaged components can become useless due to heat or more obvious physical damage that takes place when you're trying to get them off the board (such as leads so short they can't be spread enough to fit into the holes of your perforated circuit board and still have enough length to make a connection), new ones are so inexpensive that it's foolish to waste your time trying to salvage the old ones.

If you come across a surplus board full of wire-wrap IC sockets and are into wire wrapping (like caviar, an acquired taste), they are worth salvaging; wire-wrap sockets are expensive. They are frequently secured to the board with only a few dabs of solder at the corner pins, and can be worked free with relative ease.

There is a side benefit to be had from reclaiming components from surplus circuit boards. Some people knit, embroider, or prepare elaborate meals to soothe their nerves. And some find nothing more relaxing than a quiet evening coaxing components from old circuit boards and dreaming of the new circuits they'll put them into.

How to Remove Components from Boards

There are all sorts of fancy gadgets on the market for removing soldered-in components from circuit boards. Some such machines are intended for all-day, day-in-and-day-out commercial use and cost hundreds of dollars. All you need is a soldering iron and one or two other inexpensive gadgets.

The soldering iron—we'll get to soldering irons in just a few pages—can be the same one you use in your everyday construction work. The heating element should be rated at 23 watts or so. The only time you want a heavier-duty element will be for melting the solder in an area of the board where heat is absorbed very quickly—perhaps a large ground plane area at the edge or if you're working on parts that either absorb a lot of heat or are fastened to something that does. The tip on the iron, unless you normally use a very fine one with little heat-holding capacity, can be the same as the one you use to put things together.

The other item you'll need will be some solder-absorbing braid (Fig. 1-4), sold under various trade names such as Solder-Wik or Kwik-Wick. This material is a length of very fine copper wires braided just like the copper braid that is used as the outer conductor in coaxial cable. In fact, the first desoldering braid was the braid from a piece of coax. Someone undoubtedly noticed how well the braid absorbed solder during con-

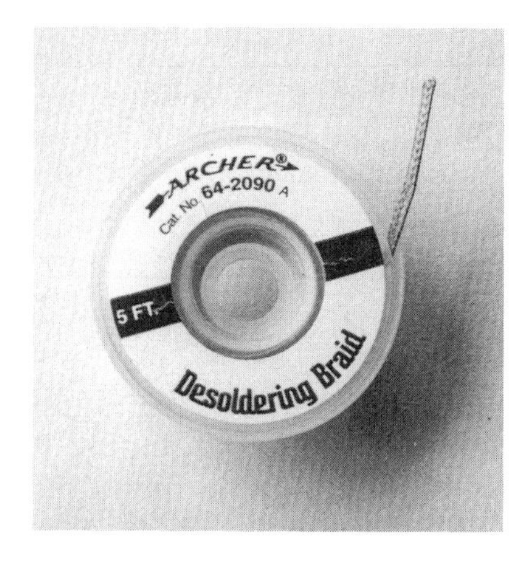

Fig. 1-4. Desoldering braid is made of finely woven copper wire. When applied to melted solder, it sucks it up like a sponge.

struction and decided to try it to absorb some excess solder from a connection.

This is how you use it. Holding one end of the braid, you place the other end in contact with the joint you want to desolder and apply the tip of the soldering iron to the top of the braid (not to the joint!). In a few seconds the copper, and the solder beneath it, will have heated sufficiently so the solder melts and is absorbed by the braid. Remove both the iron and the braid, and you'll find the joint nearly solder-free. Cut off the solder-laden portion of the braid and you'll be ready to tackle the

Radio Shack

A "solder sucker" provides the means to remove larger quantities of solder than can be absorbed by a small piece of desoldering braid.

next joint, or to go back and get the rest of the solder (there may have been too much to get all at once the first time) from the joint.

One of the qualities of copper that makes it so useful is the way it conducts heat. That's why copper is so frequently used in cooking utensils (and in soldering-iron tips, of course). Desoldering braid can rise to temperatures in excess of 600 degrees Fahrenheit and the heat can be conducted very quickly from the soldering iron up to your fingers. Be careful. You might want to hold short pieces of braid in a pair of long-nose pliers.

For situations where a lot of solder was used to make a joint, you might want to use a gadget called a solder sucker to remove most of it before going in with the braid. A solder sucker is a rubber bulb, a little like an ear syringe, with a Teflon tip that can resist soldering-iron temperatures. You use the iron to melt the solder at the joint and suck up the solder while it's still molten with the bulb. The liquid solder will cool and harden into small blobs inside the bulb. You can empty the bulb from time to time by removing the Teflon tip. With most of the solder removed from the joint, desoldering braid should finish up the job nicely. Remember to clean off any solder that stays on the tip of the iron by wiping it on a damp sponge or cloth.

So far we've described how to clean up a soldered joint, but not how to get the component out of the hole in the circuit board. Even desoldering braid will not remove all the solder from a joint, especially if a circuit board uses plated-through holes. Plated-through holes are coated on the inside, by their manufacturer, with a thin layer of metal. On top of the metal is added a thin layer of solder to make it easier to solder components to the board and to form an electrical path from a point on one side of the board to a point on the other. Double-sided circuit boards (those with circuits printed on both sides) always have plated-through holes if they are professionally manufactured, and single-sided boards sometimes have them as well.

Because the manufacturer of the circuit board did not know that you were going to take it apart into little pieces, he did not have you in mind when he plated-through the component mounting holes. Some of the solder on the inside of the holes might remain, despite your best efforts, and hold the component tightly in place.

Stubborn components can be removed by heating their leads and pulling gently with a long-nose plier. If several leads are holding a part in place, they will have to be worked on a little at a time in rotation until one, and then another, comes free. This is how a lot of breakage occurs.

Be as gentle as you can. Let the component you're working on cool off between attacks. Semiconductors and other electronic components do not get along well with heat. If they are not damaged outright by heat, the components can be aged prematurely. Try working on two or three different sections of the board alternately, or better, on two separate boards in turn. This will give things time to cool off in between rounds.

Types of Wire

Wire for electronic construction comes in two types: solid and stranded. It is also available in a number of gauges, which determine its current-handling ability.

For several reasons, 22-gauge solid wire will work best for most purposes. First, it fits conveniently into the holes in solderless breadboards (if it were stranded, you'd have a terrible time trying to get all the strands to fit into the holes!). Also, it can be bent with a pair of pliers—and will hold its bend—so it can be neatly routed from point to point in a circuit. Finally, because it is not stranded, there is no possibility of stray "hairs" causing disastrous short circuits.

The use of stranded wire generally should be reserved for high-frequency applications. A phenomenon known as the "skin effect" causes the electrons carrying high-frequency signals to prefer to travel on the outside of a conductor rather than equally through its cross section. A wire consisting of many fine filaments, then, with a lot of "outside" space, can carry more current than a similar-size solid one at high frequencies. Because it can be manipulated more easily than solid wire, you will also find stranded wire useful where heavy gauges are involved.

Wire-wrap wire is very find solid wire—about 30 gauge—used in wire-wrap construction, where solderless connections are made between specially designed square socket-pins by wrapping several turns of wire around them. The current-carrying capacity of this type of wire is adequate for low-level signals and even for carrying power to low-power IC packages. While you may not use wire-wrap construction techniques, you will find this material useful as a "wrapper" in making mechanical splices.

These same techniques can be used to remove a defective or questionable component from a circuit board so that it can be tested and then replaced with a new one. In situations such as this, the holes from which the component was removed remain clogged with solder. Solder can be removed by inserting a piece of excess resistor lead (you'll accumulate lots of these as you build; always keep a few for jobs of this sort) as far into the hole as it will go (so it's touching the blocking solder). Don't apply too much pressure or you'll bend the lead. Then, holding it with your long-nose pliers, heat the lead with your soldering iron. It will conduct the heat to the solder blocking the hole, melt it, and suddenly plunge through. If you withdraw the piece of wire immediately, you'll bring a lot of the excess solder with it. Don't wait too long or the lead will just be soldered into the hole like any other piece of wire. Hold the board up to the light to see if the hole is clear; if it isn't repeat the procedure with a fresh piece of lead. Verify that the hole is clear by sliding a piece of wire or resistor lead into and out of it; it should move smoothly.

TOOLS

It isn't necessary to have a workshop full of tools, either. Especially if you've confined yourself to working with a solderless breadboard at the beginning, you'll require only a few tools.

You'll need a couple of screwdrivers—one with a flat blade, and one with a Phillips cross-point head. These tools should be of good quality and construction, but need not be "heavy duty" ones. You're going to be working with fairly small parts, and a bit of delicacy will generally be desirable. If you can, you may want to invest in the type of tool that grips the screw and allows you to carry it to and insert it into its hole. This will save you a lot of exasperation when you're working in tight spaces. Magnetized screwdrivers can also help, and can also assist you in retrieving dropped screws, nuts, and lockwashers.

Long-nose pliers are also going to be a necessity. One pair should be as delicate as you can find, for use in bending wires in tight places and manipulating lockwashers and nuts in similarly confining circumstances. These pliers will also make a handy retrieval tool for dropped parts that wind up w-a-a-y down in an otherwise inaccessible spot.

A second, heavier pair of long-nose pliers may be useful in holding nuts and other parts still when tightening up things. The more delicate tools might not be rugged enough or provide a good enough grip. They're more for bending than for holding onto things.

Little Lost Parts

Here's a trick for getting screws and small bolts into holes in tight places. Dip the blade of your screwdriver into rubber cement and glue the head of the screw to it. The cement will take only a few seconds to dry and you will then be able to maneuver the part down to where you want it, and start it in its hole. Once it's in place you can tighten the screw normally. Don't dribble rubber cement down into your work, though. It's flammable even when dry. If you put some rubber cement on both the screwdriver tip and the screw, and then let it dry before bringing the two together, the grip will be even better.

A screwdriver tip with a blob of rubber cement on it may also work as a last-resort retrieval tool for parts that have been dropped into otherwise inaccessible places.

If a wire cutter is built into your pliers, ignore it. You'll do better with cutters designed specifically for the purpose. Wire cutters are sometimes referred to as "diagonal cutters" or "dikes." Again, because you'll be working on a rather small scale, you won't want the heavier-duty kind found in most "household" tool boxes. Get a pair of the smaller ones intended specifically for electronics use, spring-loaded, with slightly angled jaws, for nipping off excess lead lengths right down to the joint.

You'll also need a wire stripper. Rather than one of the big bulky ones with holes of various sizes, you'll be better off with one of the smaller, adjustable ones. The size of the stripping aperture is usually changed by rotating an offset disk that stops the jaws of the stripper from closing at a certain point. Use a short length of the wire you'll be working with to adjust the tool. Keep closing the size of the stripping hole until the tool removes the insulation from the wire and leaves a very slight nick in it. Then back off, just a bit, so that the wire is left unnicked. If you nick a solid wire, especially a fine one, it may break later after it has been subjected to some amount of stress and vibration. And if you nick a stranded wire, some of the outer strands may be severed completely and find their way into places they have no place being—as in creating a short circuit between two points. Don't try to use this sort of tool for working with wire-wrap wire; it will just frustrate you. There are special strippers—frequently built right into wire wrap tools—specifically for this purpose.

Wire wrapping is not for beginners, but can become an acquired taste as your skills develop. This wire-wrap tool is the size of a small screwdriver. The small chisel-shaped object is a wire stripper, which is sometimes built right into the handle of the wrapping tool itself.

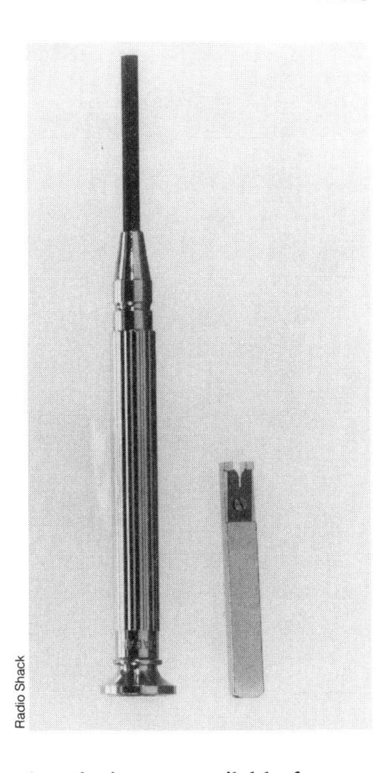
Radio Shack

Although precut and prestripped lengths of wire are available for use with solderless breadboards, they're generally not worth bothering with. You're better off cutting your own wire to length. Not only will you get exactly what you need, but you'll save money as well.

A volt-ohmmeter (VOM) like the one shown in Fig. 1-5 is a good investment. You'll want if for troubleshooting and testing components. While a digital multimeter (''multimeter'' is another name for a VOM) of the sort shown in Fig. 1-6 will generally give you more accurate readings, the analog meter on a conventional VOM will give a faster indication of continuity. This is a frequent indication of whether or not a component is good or whether a short or open circuit exists.

The ''continuity'' setting on most digital meters, which causes a tone to sound if current passes between the test probes, can be used to detect short circuits, but it takes too much time to respond. If you're in a hurry or are running short on patience as you make your tests, you might mistake this time lag for an indication of an open circuit. The pointer of an analog meter, however, will start to swing immediately.

Get the best meter you can afford. You will need a good meter not so much for added features as for accuracy, at least in analog meters. Or

Fig. 1-5. A VOM such as this one with an analog meter may not be quite as accurate as a digital one, but it can be more useful for quick readings, especially where continuity tests are involved.

maybe you can invest in a digital meter for accuracy and an inexpensive analog one just for continuity tests.

If you're feeling affluent and intend to do a lot of hardware work, a set of nutdrivers might come in handy. Indeed, some people consider them a necessity. And, if you're working with perforated construction board, you'll undoubtedly want a small hacksaw or coping saw to cut big pieces into smaller project-size ones. For preparing metal chassis and cabinets, an assortment of files and drill bits, as well as other metal-working tools, will come in handy.

Power

By now the question has probably arisen in your mind: "How am I going to run the stuff I build? With my Radio Shack battery card?"

Fortuitously, the first of the "project" chapters in this book deals with power supplies, and you'll be able to build just what you need to run the rest of your projects. Of course, you could use batteries and, in fact, might want to build the final version of some of the circuits described specifically for battery operation.

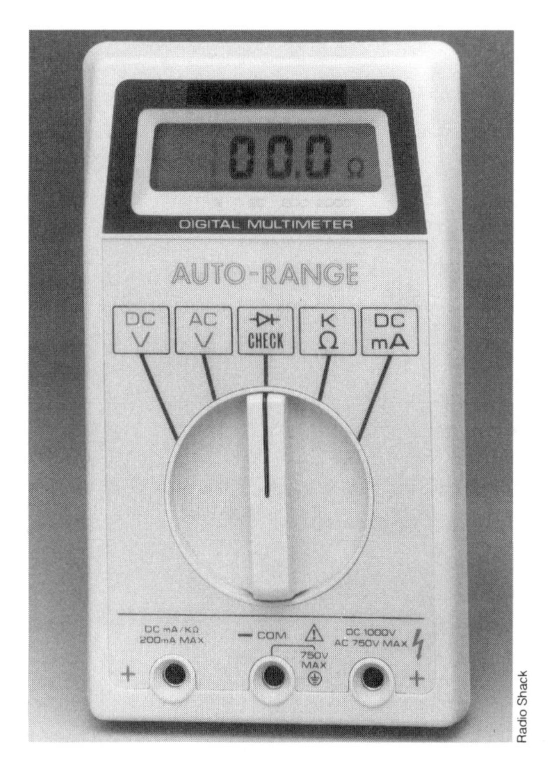

Radio Shack

Fig. 1-6. A DMM, or digital multimeter, is a relatively inexpensive way to obtain good accuracy in your measurements.

For now, though, you can build one of the supplies described in Chapter 8 or, taking the coward's way out, buy one. If you're going to do that, get a supply that will provide about 1 ampere at 5 volts dc and plus-and-minus 12-15 volts at a hundred milliamperes or so. The positive-negative supply will be for use with operational amplifier circuits described in Chapter 11.

Depending on your applications, having separate supplies might prove less expensive. Even if you're not going to build your own power supply (or supplies) read Chapter 11 before you make your purchase. It will help you to understand what to look for.

Other Tools

There are lots of other little things you'll come across both in this book and in your wanderings through stores and catalogs in search of parts and tools. Some items will have lasting practical value; others will probably turn out to be disappointments and wind up in your junk box.

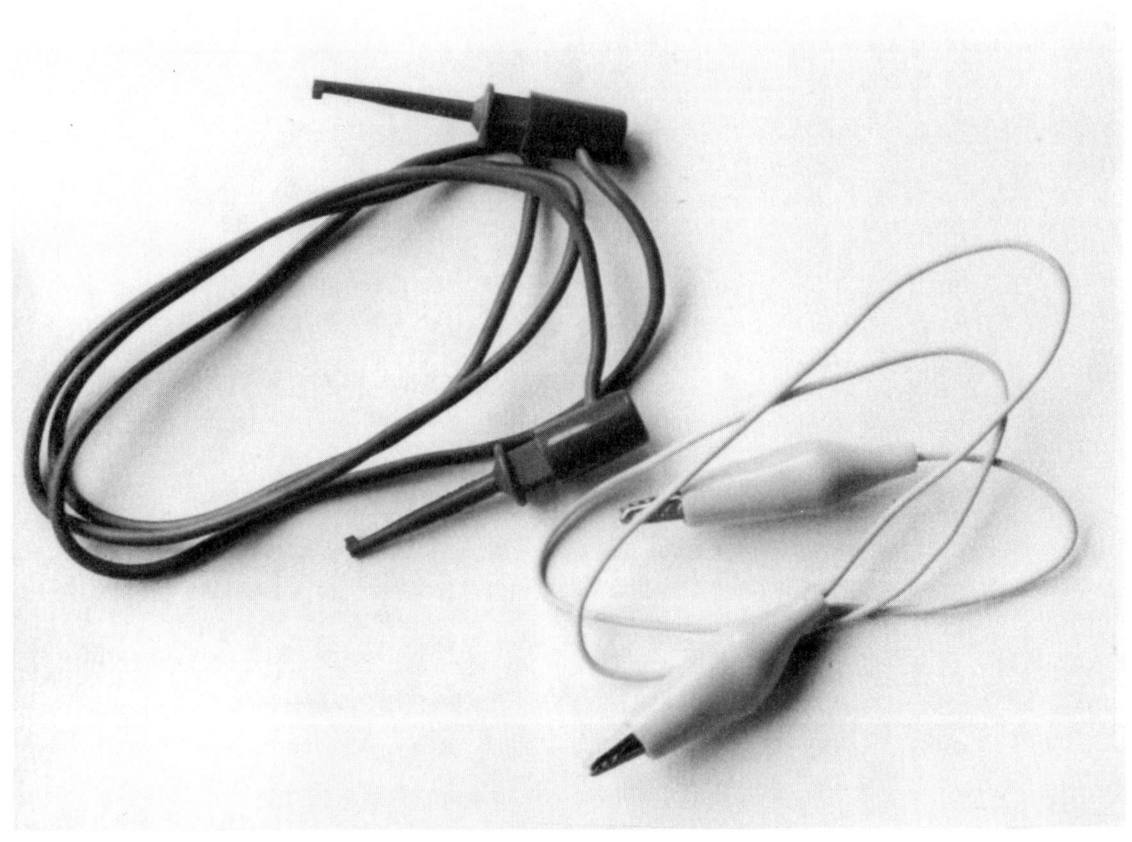

An assortment of test leads with different types of clips will prove useful. The miniature micro clips (top) are invaluable for making temporary connections to IC pins or to other component leads. They also eliminate the risk of shorting two IC pins together as might happen with alligator clips (bottom).

It's impossible for someone else to tell you what you're going to like or dislike. If some gadget appeals to you and you can afford it (or somehow justify its expense), then get it. If it doesn't work out, relegate it to the junk box where you may find a use for it later or where it may eventually be discovered by someone who really needs it.

2
Soldering Equipment and Techniques

This is what you've been waiting for, right? The *real stuff*. Soldering! This is what makes putting together electronic circuits different from building other things like model cars, and is where we separate the men from the boys.

IRONS

Let's begin with the heavy artillery, the soldering iron. There are all sorts of irons, and they come in all sizes, shapes, and varieties. Remember, as you choose your tools, that unless you're building something big and heavy there is a certain degree of delicacy about the work you're going to be doing. Your soldering iron should not be big and heavy; indeed, much of what you look at as you decide which to purchase will be too clumsy for your purposes. What you want are the following: instant heat (but not too much of it), small size, and ease of use.

That rules out those big, plumber's-type irons, propane- and butane-fueled ones, and soldering guns. They either generate too much heat, or will simply overwhelm your work with their size and awkwardness. Furthermore, soldering guns, sometimes called "pistols," require time to come up to temperature; every time you let go of the trigger they cool down.

Cordless irons, which operate from rechargeable nickel-cadmium batteries, have their good points and their bad ones. On the positive side, they're pretty convenient to use, especially since their lack of a trailing wire makes them easy to maneuver and their plug-in heating-element tips are small and are not likely to fry a board.

On the other hand, when you use them you have to hold a trigger button with one finger. You may find that this somewhat restricts your dexterity. It also takes several seconds after you depress the trigger for the element to come up to working temperature. Finally, unless you make extensive use of such an iron—every day or two—the longevity of its battery is going to suffer. Nickel-cadmium batteries need to be worked hard and often to keep them in shape, and if they just sit idle, or in the charger, they will fall into bad habits and refuse either to charge fully or to hold a charge for as long as they should. Maybe you should

Radio Shack

Fig. 2-1. A soldering station puts it all together: iron, wiping sponge and, usually, thermostatic control of tip temperature. Frequently the tip of the iron is grounded to prevent damage to delicate ICs by static electricity.

hold off on one of these for a while, until you know what your usage is going to be.

A soldering station such as that shown in Fig. 2-1 is a combination of a soldering iron with a stand and, usually, a pad for cleaning the iron. The iron usually comes with a thermostatic device intended to keep the tip temperature relatively constant, and is also grounded to prevent static electricity buildups that might damage sensitive components on which it is used. At this stage of the game, you don't need something this elaborate.

Your best bet to start with is a plain old plug-in soldering iron or pencil with interchangeable heating elements and tips (Figs. 2-2 and 2-3). For most work you'll want a tip rated between 15-23 watts or so; this will be perfect for just about all your circuit construction work. You may want to have a second element, rated at perhaps 30 watts or a little more on hand for work in areas or with components that soak up a lot of heat. This large-capacity element can also prove useful should you have to use it out of doors, where wind can suck away a lot of the heat that should be going into melting your solder.

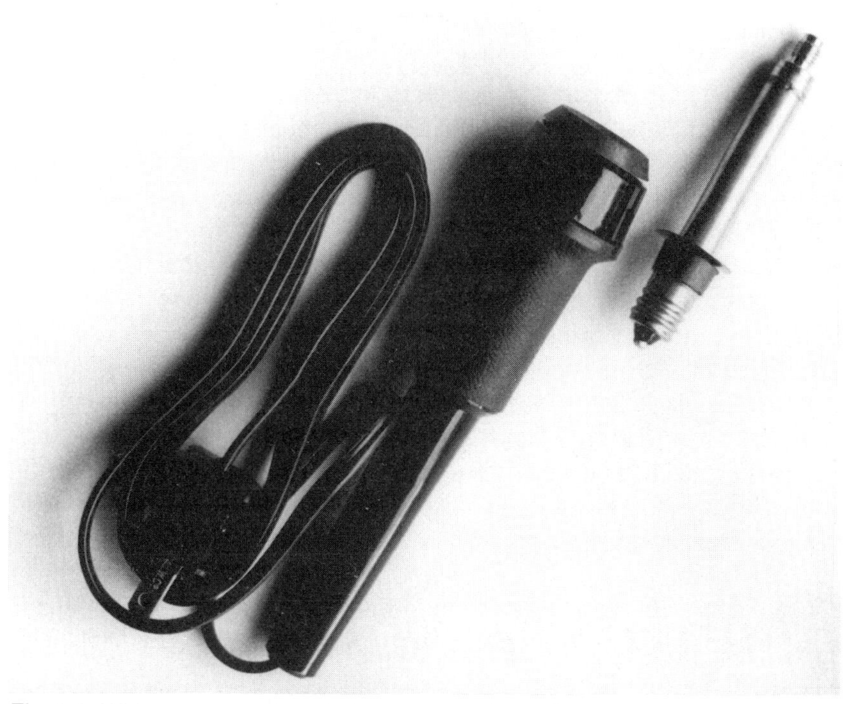

Fig. 2-2. With a soldering iron that uses screw-in heating elements you can select the heat most appropriate to the job at hand.

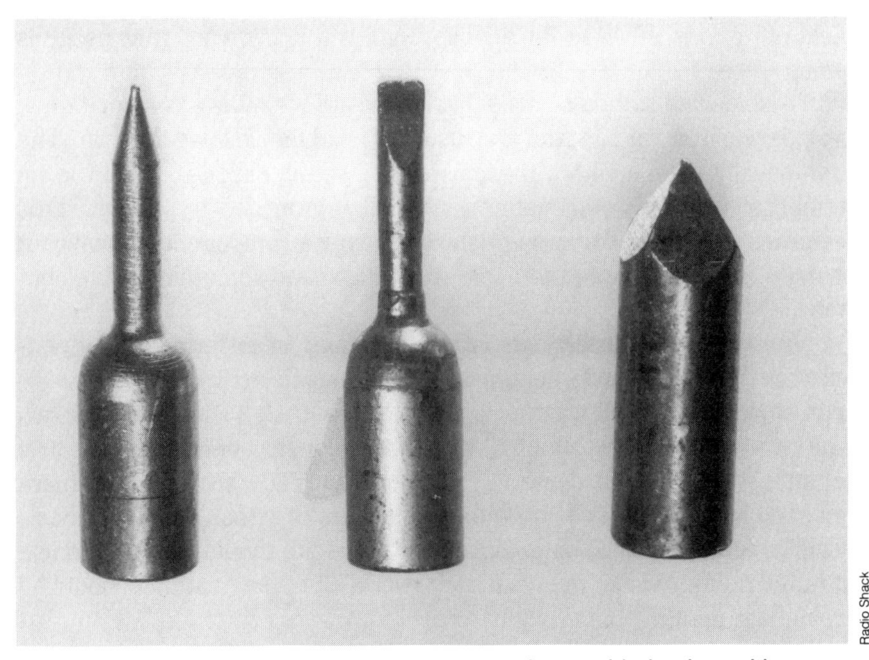

Radio Shack

Fig. 2-3. Interchangeable tips allow you to customize a soldering iron with screw-in heating elements (Fig. 2-2).

Screw-on interchangeable tips turn out not to be very interchangeable. Once they've been used, the heat seems to bind the threads and the tips are more or less permanently in place. You can sometimes remove a tip that's frozen on by using two pair of pliers, one to hold the tip and the other to hold the heating element onto which it is screwed, but you risk damaging the element in doing so. Better select a tip-and-element combination and then use it consistently. If you need a second one, use a second tip-and-element set.

The best and most durable type of tip is the steel-plated copper one. For circuit board work a 1/16 inch chisel tip of this type in combination with a low-wattage element can't be beat. Not only is it difficult to apply too much heat with this combination, it's also a deterrent to drowning the joint in solder and creating solder bridges. You are naturally constrained from going overboard by the size of the tip.

The slightly larger general-purpose tips are also good, although those that are unplated copper tend to wear down with use. Their shape can be restored by gentle filing or sanding, followed by a retinning. Round or chisel points work well.

Avoid, however, the big chisel-point copper tips. They're intended to store (and release) a lot of heat. Besides being awkward, if not impossible, to use because of their size, they're liable to fry the board you're working on, lifting traces and possibly ruining the components you're working with.

HOW TO SOLDER

The big trick to soldering is "don't use too much . . ." Don't use too much solder, don't use too much heat, don't use too much time. If you keep this in mind and, more important, practice it, you'll be doing a professional job practically from the start.

Almost any solder you can find in a store that sells electronics supplies will be suitable; avoid products such as silver solder—they're for special purposes that have no business here. And so-called liquid solders, or those that come in little flat pieces that you're supposed to melt by holding a match to them, they are a joke. What you want is a 60/40 rosin-core solder. The "60/40" refers to the percentages of the metals making up the solder: 60% tin and 40% lead. More tin and less lead makes a harder solder with a higher melting temperature, and reversing the proportions lowers the melting point and makes a softer material. 60/40 is about right for general purpose work.

The rosin core is what's called a flux, and the solder you get may be labeled "rosin flux." "Flux" is Latin for "flow," and that is the purpose of the flux, to make the solder flow better and thus "wet" the materials it comes into contact with. Just as on a really clean piece of glass, water does not bead but, rather, tends to spread out and form a sheet or film, so should solder flow over a joint. If it doesn't, but forms balls and blobs instead, the materials are dirty and have to be cleaned. Flux helps, but it is not a substitute for starting with clean materials.

You may occasionally run across acid flux, in solder or by itself in a tin. *Avoid it like the plague!* Acid flux, which is used mainly in the construction and plumbing trades, is corrosive and has no place in electronics work.

While the diameter of the solder is not critical, for delicate work on circuit boards—for instance, a smaller diameter, 22 gauge, will work better than a larger one. You'll be able to control the placement and flow of the solder more closely, and its small diameter will make it more difficult for you to use too much.

Cleanliness is more important than you may think. Usually, it will not be a problem—you'll probably be working with new components, freshly stripped wire, and so on. From time to time, though, you may run into a situation where solder will not adhere to the material it's supposed to. Putting aside for the moment the possibility that you're trying to solder something that won't accept solder (aluminum, for example), the problem is probably dirt or some form of oxidation.

If there's dirt or some other residue on a wire or component lead—maybe leftover adhesive from a piece of electrical tape from when the component was last used—it can generally be removed with alcohol (90% isopropyl, available in drug stores) or with acetone. Acetone is the main constituent of nail-polish remover, but is available in pure form in hardware stores. Be careful in using these solvents, particularly acetone. Acetone will turn many plastics into goo! It's also extremely flammable; don't smoke around it.

You may run across integrated circuits whose leads are so oxidized that they won't take solder or make good contact when inserted into sockets. These can be cleaned up by dipping them in a tarnish-removing solution such as Tarnex, and then rinsing them with water and drying them well.

There is a mechanical aspect to the art of soldering, too. That is, solder should not be used as a glue. Don't trust two pieces of wire to hold together simply by virtue of a drop of solder applied to their junction. There should be mechanical integrity to the joint as well.

Sometimes, as is the case with IC sockets inserted into a printed-circuit board, it is not possible to make that kind of good mechanical contact between the pins of the socket and the traces of the board, but considering the insignificant weight of the socket and the large number of connections involved, this is not a major problem. In point-to-point wiring, though, a good mechanical joint is almost a necessity.

When you are working on circuit boards and inserting devices such as transistors, capacitors and resistors which have fairly long leads—at least if they're new—after inserting the leads into the appropriate holes you can bend them in opposite directions both to hold the components in place temporarily, and to lend some mechanical integrity to the joint. Sometimes, though, all you can do is twist together the wires you'reconnecting.

OK, here's how to solder. Apply the tip of the iron to the joint so it contacts all the pieces that will be involved in the connection. After allowing several seconds for heat from the tip to be transferred to the joint (exactly how long depends on several factors—the wattage of the heating element and the mass of the materials being soldered, to name two), apply the end of the string of solder to the point where the iron and the other materials meet. Do not apply the solder to the tip of the iron alone and expect it to flow down over the joint; nor should you apply it just at the joint—there may not be quite enough heat there.

How much solder is enough? The solder should just flow over the parts to be joined and wed them gently; a large ball of solder is of no use and, in fact, may be detrimental to your success. You should apply heat and solder just until the solder begins to flow—that will usually provide enough material to do the job.

Too much solder can do more than just hide improperly soldered joints. A big blob of solder can be an indication that the joint is dirty and that, underneath the blob, the solder has not flowed as it should have. Solder blobs can also be responsible for short circuits on a circuit board.

Most professionally manufactured circuit boards are coated with what's called a solder mask. This is usually a greenish material that covers all of the solder-side of the board except for the small areas around component holes, where solder is to be applied. Solder won't adhere to the material of the mask, and it protects much of the board (such as the signal- and voltage-carrying traces) from inadvertently being shorted out by solder drops and splashes.

Even with this mask, however, it is still possible for a heavy-handed solderer to create a blob, shorting out two adjacent solder pads. Such blobs can sometimes be made to "disappear" by reheating them. The excess solder might migrate to one or the other of the points it accidentally connected, or wind up on the tip of the soldering iron, from which it should be removed. Desoldering braid can also be used to absorb and remove the excess, although the joints might have to be resoldered if this is done.

There is also a possibility of a cold-solder joint, caused by not enough heat being applied to the joint, or by the heat being removed too soon. A cold-solder joint can usually be recognized by its "blobby" appearance, and a grainy surface. A proper joint shows evidence of the solder's having flowed smoothly and is somewhat shiny, although you shouldn't expect a mirror finish. Cold-solder joints can usually be fixed by reheating them until the solder flows. It might be necessary to remove the old solder (with desoldering braid) and then apply fresh solder.

TINNING AND CLEANING

The tip of a soldering iron has to be tinned, before you ever use it, or if you ever file or sand it down to bare copper in reshaping it. Tinning means applying a coat of solder to the tip and letting it "cook in" for a minute or two.

Tinning is necessary, because the bare copper of the tip will quickly oxidize when heated, and the oxidized surface will repel solder, making it impossible to heat the solder properly and get it to flow the way it should.

For this reason, the first thing you should do with a new tip is to tin it. "Wet" the tip thoroughly with solder and let it cook in for a minute or two. Then wipe the excess solder off on a cleaning sponge or damp paper towel (you can shake the excess off first onto a piece of newspaper).

Speaking of wiping off the tip, you should always do this after you have finished soldering. Not after you have finished for the day, but after every joint (or series of joints if you're doing a dozen or more in a row). A tip on which excess solder is allowed to remain will quickly develop an oxidized "crust" that will greatly reduce the efficiency of the iron, if it doesn't make it impossible to solder at all. A clean soldering-iron tip should look slight wet with solder. Use a damp cleaning sponge (which you can get at the same place you purchase your iron and other supplies) as shown in Fig. 2-4, or even a damp wad of paper towels. Whatever the wiping material, it doesn't have to be wet, just damp. Keeping the tip clean will prevent it from oxidizing and will make your soldering easier, and the joints more reliable.

The last item you should have is a stand to hold the iron when you don't have it in hand. While you could use a heavy ashtray, the heat-sink-

Fig. 2-4. Use a dampened sponge or paper towel as a wiper to remove excess solder from an iron's tip. Doing so will extend the life of the tip.

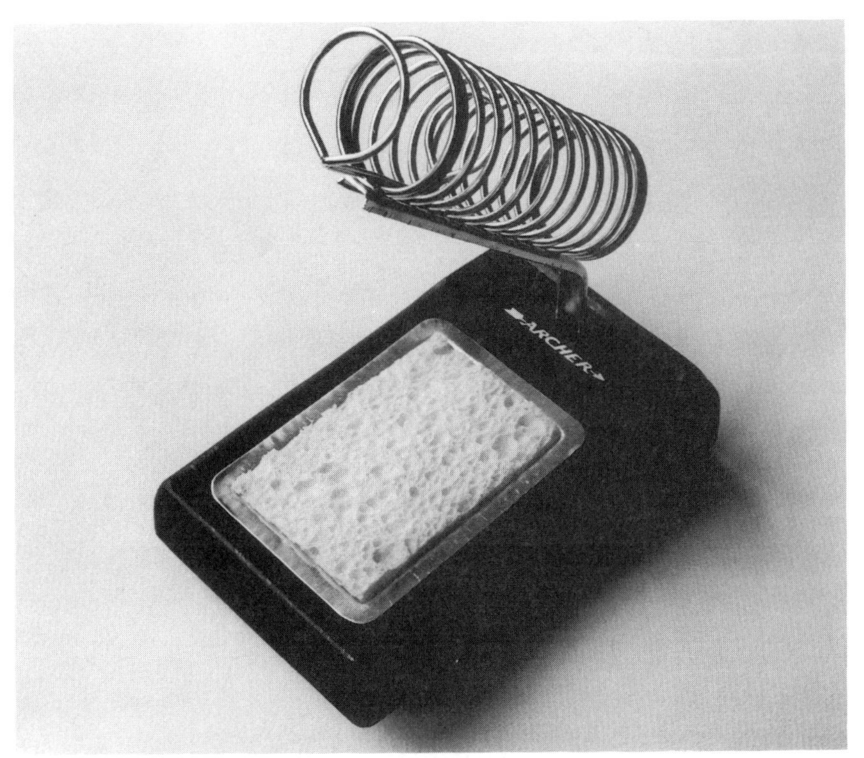

Fig. 2-5. A stand, with or without a built-in wiping sponge, provides a place to keep a hot soldering iron ready for use and out of harm's way.

ing properties of a stand such as that shown in Fig. 2-5 will prolong the life of the tip. It will also lessen the chance of someone's bumping or brushing against the iron and getting burned (or knocking it onto something that can be burned). It's cheap insurance.

3
Ohm's Law and Other Mathematical Stuff

Building things is fun. Unfortunately, you can't just put together a device by throwing in a little of this and that, the way your mother used to make a pot of soup. You can't—until you've been doing it for years and years—work without some sort of plan beforehand.

This does not have to be a full schematic, with all the parts symbols drawn in and neatly labeled, but you do need to know what type of components you're going to use, and what their values are going to be. And this means that in many cases you're going to have to calculate those values. Well, you didn't think the numbers were going to just pop into your head, did you? You're going to have to work a little for them!

To many people, the thought of this part of electronics work—as opposed to the fun part of stuffing circuit boards and soldering things to each other—is terrifying, and is enough to make them think seriously of turning to some other pastime such as gardening or potlatching. Determining component values mathematically needn't be that much of an ordeal, though. There's no way of making solving equations as pleasant as eating a piece of chocolate cake, but working with numbers is nothing to be afraid of, either.

IT ISN'T REALLY ALGEBRA

X, the unknown. That's what algebra may have been to you. When you were in high school, maybe just the thought of it was enough to start you quaking. If you had the choice, you probably elected not to take it back then.

Well, your time has come! What we're about to discuss is technically algebra. Not the really fancy kind with lots of *x*'s and *y*'s and square roots and Greek letters, but it is a form of algebra nonetheless.

All algebra really is, is a mathematical procedure for finding the value of something you don't know by using values that you do know. And, when you're trying to determine what value of resistor, for example, you need in a particular place, that's exactly how you make that determination. You use what you do know to find out what you don't.

OHM'S LAW

One of the fundamental relationships in electronics is stated by Ohm's law, named after G.S. Ohm, the German scientist who formulated it. In fancy language, Ohm's law says that in an electrical circuit, the voltage drop is equal to the value of the current multiplied by the resistance of that circuit.

An analogy is frequently made between the flow of electricity through a wire or circuit and the flow of water through a pipe. The volume of the water represents the current, the water pressure is the voltage, and the diameter of the pipe is the resistance.

E, I and R

You would think that in electronics formulas and equations values would be represented by letters that reminded you of what they stood for. Volts would be represented by "V," amperes by "A," and resistance by "R." And, in many of the formulas found in data books this is the case . . . but not always.

Ohm's law is an excellent example of the exception to what logic tells you should be the rule. Voltage is represented as "E," and amperes by the letter "I." Resistance, fortunately, gets an "R." But where did the "E" and the "I" come from?

In all probability, "E" stands for "electromotive force," or in plain English the force, or pressure, that causes electrons to move. In other words, voltage. The letter "I" most likely represents "instantaneous current," or the electronic current (amperage) at a given instant.

Knowing the origin of these abbreviations may not help you in your equation solving, but it will save you a lot of time that would otherwise be spent wondering where they came from.

A wide pipe that offers little resistance to the flow of water can handle a high volume of water (current) at low pressure (voltage). Similarly, this same pipe will have the same output if the current, or rate of flow, increases, while the pressure (voltage) is reduced. And, if the pressure (voltage) is increased while the diameter of the pipe remains the same, the amount of water flowing through the pipe (current) will also increase. Changing the resistance, represented by the diameter of the pipe, will affect the other two.

That's what Ohm's law says if the concepts of water and plumbing are substituted for electricity and wire.

Mathematically, Ohm's law is expressed as:

$$E = I \times R$$

This is about as deep into algebra as we're going to get. Before we discuss a few of the uses of Ohm's law, let's look at the way the equation works. These principles apply to all equations.

More About Algebra

The values represented by "E" and by "I × R" are separated by an "equals" sign (=). What that sign indicates is that the value on its left, E, is *equal* to the value on its right, the value calculated by multiplying I by R. In fact, it is this equality that gives an *equation* its name. Algebra might be called "the mathematics of equations." There is always an equals sign in an algebraic formula, telling you that what's on the left side of the equation is equal to what's on the right.

Now, if you know what's on one side of the equation, you can easily find out what's on the other side, since you know that the two are going to be equal. In the case of Ohm's law, the voltage drop is equal to the current multiplied by the resistance. If the current were 2 amperes, and the resistance were 5 ohms, multiplying them would give a product of 10 (2 × 5), and the voltage drop would be 10 volts. That's all there is to algebra! Well . . . almost.

Since the values of both sides of an equation are the same, if we do the same thing to both sides, they'll still be the same. Suppose, for example we were to divide both sides of the equation by I. That would give us an equation that read:

$$E \div I = R$$

What happened to the I on the right side of the equation? Well, when you divide something by itself, the result is always one. Go ahead, see for yourself:

$$5 \div 5 = 1$$
$$144 \div 144 = 1$$
$$1 \div 1 = 1$$
$$I \div I = 1$$

And, anything multiplied by 1 is still just itself:

$$5 \times 1 = 5$$
$$144 \times 1 = 144$$
$$1 \times 1 = 1$$
$$R \times 1 = R$$

The rearranged equation tells us another truth: that R is equal to $E \div I$. Now we can use Ohm's law to find both voltage and resistance.

Finally, by dividing both sides of the original equation ($E = I \times R$) by R, we can see that: $E \div R = I$. Current (I) equals voltage (E) divided by resistance (R). That's how Ohm's law works.

Ohm's Law Simplified

Now that you know the "complicated" form of Ohm's law, here's the simple form. Figure 3-1 shows the three letters, used in Ohm's law, "E," "I," and "R," arranged with the "E" above the other two and separated from them by a straight line. Let's say you want to determine the current flowing through a resistor. If you cover the letter "I" with your finger, you'll see that what you can still see is "E ÷ R." And that's how you find the value for the current—divide "E" by "R." Similarly, if you cover the "R," you'll see that it's equal to "E ÷ I." And, finally, when you cover "E," the result is "I × R." All you have to do is single out the letter whose value you want to know, and use the relationship between the other two to find it.

$$\frac{E}{I\ R}$$

Fig. 3-1. Ohm's law. Cover the "unknown" you are looking for with your finger and you'll find the relation of the other two variables you need to solve for it.

Using Ohm's Law

Because you're interested primarily in determining the values of the components you are going to use, most of the time you'll use Ohm's law in the form "R = E ÷ I" to determine the values of the resistors you need.

Let's look at just about the simplest circuit there is, a resistor in series with a battery. It doesn't do anything useful except run down the battery and heat up the resistor, but it does serve nicely to illustrate Ohm's law. For our purposes, this particular battery outputs 12 volts. Let's say that you want to draw a current of 2 amperes from this battery through a resistor.

The circuit is not going to draw 2 amperes just because you want it to. It is the resistor that's going to determine how much current flows. The question is, what value resistor do you need? Here's where you can use Ohm's law.

You know that R = E ÷ I. And, in this case, you know that E is 12 volts, and that you want I to be 2 amperes. The equation therefore becomes:

$$R = 12 \div 2$$

or

$$6 = 12 \div 2$$

You need a 6-ohm resistor to make the circuit function the way you want it to.

In some circuits you will want to use a resistor to get a certain voltage, let's say to get 3 volts from a 5-volt supply. You will do this by passing the current to the point that needs the three volts through a resistor. The question, of course, is what value will that resistor have to be?

Now, you know that you're going to want to use Ohm's law to find the value of the resistor, buy what are you going to use for the values of E and I? You have a couple of possible values for E (don't jump to any conclusions, there will be more about this shortly), but what about I?

Generally, in situations such as this, where a so-called dropping resistor is used, the current, I, will be a constant (there are, however, situations where it will vary). And, perhaps your eventual target is a power consumption of so many watts. We'll come to this shortly, too—many units of electrical measurement are interrelated, and it's just as difficult to talk about them independently as it would be to explain them all at once—so for the moment let's just assume that the current drawn will be 0.5 amperes. (It's always a good idea to include the zero before the decimal point. This shows that you intend the decimal point to be there, and that the number is not, perhaps, 5 amperes with a flyspeck in front of it.)

Good, now you have I, but what's E: 5 volts or 3 volts? The answer is, neither. The figure you need to use in Ohm's law here (and everywhere else, for that matter) is the voltage *drop* across the component or components in question. That is, the difference in electrical potential between the starting point—the 5-volt end of the resistor in this case—and the ending point—the 3-volt end.

Now it becomes clear! The voltage drop is 5 − 3, or 2 volts. Now you have all you need to use Ohm's law to determine the value of the resistor:

$$R = E \div I$$
$$R = 2 \div 0.5$$
$$R = 4$$

Ohm's law triumphs again!

OTHER UNITS OF MEASUREMENT

So far we've dealt with three units of electrical measurement: volts, ohms and amperes. These are used to represent, respectively, electric potential, electric resistance, and electric current. There are several other units of measurement with which you will have to deal as you design and redesign circuits.

Watts Up, Doc?

Returning to the water-pipe analogy, what about all that water that comes out the end of the pipe? If that water were electricity, the water that would be caught in a bucket at the end of the pipe would be called *watts*. The watt is electricity's measure of power and is defined as the product of voltage and current. In other words:

$$P = E \times I$$

One watt of power equals one volt times one ampere. Five watts can be 5 volts times 1 ampere, 1 volt at 5 amperes, or any other combination of voltage and current whose product is 5.

If we know any two of the three values, we can find the third. For example, suppose you wanted to build a power supply capable of delivering 30 watts of power at 5 volts dc. You would need to know how much current you have to be able to provide so you could be sure of using components that could handle the load. And, since you are now an expert on Ohm's law and algebra, the solution to this should be easy. If you plug the values of 30 watts and 5 volts into the equation for wattage, and then divide both sides by E, which is 5, to isolate the unknown, I, you get the answer:

$$W = E \times I$$
$$30 = 5 \times I$$
$$30 \div 5 = I$$
$$6 = I$$

You'll be dealing with a current of 6 amperes.

Occasionally you might find yourself in situations where you know you'll have to use Ohm's law, but have only one of the three values you need, along with a value in watts. Well, you can't use the wattage directly in the Ohm's law equation, but you can probably break it down into its voltage and current components, and then one or the other of these will be the missing figure you need to make Ohm's law work.

Capacitance

One of the most commonplace and versatile electronic components is the *capacitor*, and we'll spend some time talking about the different types of capacitors, and which applications call for which types, but that will be later. For the moment, we'll just talk briefly abut how the values of capacitors are measured.

A capacitor is a device that is used to store electrons temporarily. Batteries—even so-called storage batteries—generate electrons as the result of a chemical reaction. Capacitors merely store them.

The basic unit of capacitance is the *farad*. A capacitors's capacity to hold electrons is measured in farads. The farad was named in honor of the scientist Michael Faraday, an early electrical experimenter responsible for a number of discoveries. A farad is defined as one coulomb of electricity applied at a potential of one volt. That is no trivial unit, since a coulomb contains 6.25×10^{18} (6.25 thousand million billion!) electrons. That's a lot of electrons! Fortunately, we won't be working with coulombs here so you don't have to know more about a coulomb than that it defines a quantity of electrons.

A farad is such a huge quantity that until recently capacitors with values on the order of a farad were special-order items used principally in applications such as sub-atomic-particle accelerators where huge amounts of energy had to be stored. Within the past few years miniature devices with capacitances of 0.5 farad, one farad, and more have become available. They are frequently used to provide standby power for low-power memory ICs.

The capacitors used in most electronic devices are considerably smaller than this, both in size and in capacitance. In fact, the capacitances required in most applications are so small that capacitors are usually rated in microfarads—millionths of a farad. Often, even that is too large and capacitances have to be measured in picofarads. A picofarad is a millionth of a millionth of a farad and you may sometimes find it referred to as a micro-microfarad.

The point of all this is to warn you to watch out. Some formulas, such as those that are used to calculate the time constants of various circuits (see Chapter 9), require resistance and capacitance values. And, while the figure for resistance uses ohms, the same way resistors are marked, the figure you supply for capacitance must be in farads, not in the micro- or picofarads that define those component values. If you don't provide a number of the correct order of magnitude, your answer will be off by a factor of a million (or a trillion if you use picofarads instead of microfarads). When the time comes to use these formulas, we'll remind you of this, and show you how to juggle the numbers with ease.

RESISTORS AND CAPACITORS IN SERIES AND PARALLEL

The terms "series" and "parallel" are used to describe the way components in an electronic circuit are connected to one another. The resistors in Fig. 3-2*A* are connected in series—one after the other. Those in Fig. 3-2*B* are connected in parallel. Similarly, the capacitors in Fig. 3-3*A* are in series, and those in Fig. 3-3*B* are in parallel.

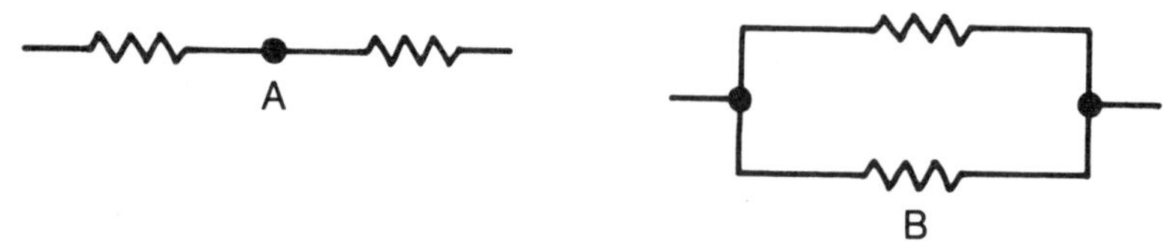

Fig. 3-2. When resistors are connected in series (*a*) their total resistance is the sum of their individual resistances. When they are connected in parallel (*b*) the total resistance is less than that of any single resistor.

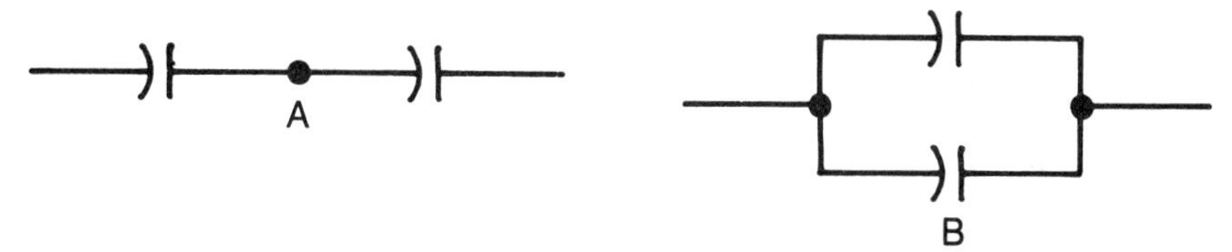

Fig. 3-3. In the case of capacitors, the combined values of devices in series (*a*) and in parallel (*b*) are calculated in the opposite fashion from those of resistors (Fig. 3-2). Capacitors in series have a total value that's less than that of any single one; capacitors in parallel form the equivalent of one bigger capacitor whose value is equal to the sum of the individual capacitances.

In a series circuit, all the current must flow through all the components. In a parallel circuit, it is divided up in proportion to the values of the individual components. Again, the water pipe analogy can help in visualizing what goes on. The total resistance of a parallel circuit is less than that of any of the individual branches.

To calculate the total value of resistors in series, you simply add their values:

$$R_T = R1 + R2$$

The flow of current through resistors connected in parallel is inversely proportional to their values; that is, more current flows through the resistor with the lower value. The equation for resistors in parallel is:

$$\frac{1}{R_T} = \frac{1}{R1} + \frac{1}{R2}$$

or

$$\frac{1}{\frac{1}{R1} + \frac{1}{R2}}$$

If there are only two resistors involved (sometimes there may be more), a simple form of the equation may be used:

$$R_T = \frac{R1 + R2}{R1 \times R2}$$

Note that this works *only* for situations where two resistors are involved.

In the case of capacitors, the situation is reversed—when you connect them in parallel, all the plates of the capacitors, which is where electrons are stored, are connected together to form the equivalent of bigger plates. The total value of capacitors connected in parallel is the sum of their individual values, while the total value of capacitors in series is less than that of any of the individual values. In parallel:

$$C_T = C1 + C2$$

while in series,

$$C_T = \frac{1}{C1} + \frac{1}{C2}$$

or

$$C_T = \frac{1}{\frac{1}{C1} + \frac{1}{C2}}$$

These fractions look more intimidating than they actually are. A little pocket calculator easily makes them manageable.

4
About Components

As you use Ohm's law and other formulas to calculate the values of the resistors and capacitors you'll need, you're going to come up with some pretty strange values: maybe 37,400 ohms, or 1.356 microfarads. You'll quickly discover that these values do not appear in the lists of standard component values provided by manufacturers or vendors of these parts.

STANDARD VALUES

Indeed, if you scrutinize a listing of standard values for resistors or capacitors, you'll soon notice that, although the values grow larger as you go down the list, the prefixes to these numbers—the digits before the zeros—start repeating very quickly. Table 4-1 shows the resistor values that might be available from a mail-order vendor of electronic components. Notice how most of the base-values repeat. The same numbers appear over and over again, followed by larger and larger numbers of zeros indicating larger and larger values.

Similarly, Table 4-2 shows a listing of standard capacitor values. Again, the same numbers repeat themselves time and again. And, if you compare the numbers in Table 1 with those in Table 2, you'll see that many of them are the same, or nearly so. Compared to all the possibilities, there are only a few values of resistors and capacitors manufactured. There aren't any values of the sort you'll come up with again and again in your calculations—where are you ever going to find a 1.356-microfarad capacitor, or that 37.4k resistor?

Table 4-1. Standard Resistor Values.

10	180	1.2K	22K
15	220	1.5K	33K
22	270	2.2K	47K
33	330	3.3K	100K
47	390	3.9K	150K
68	470	4.7K	220K
82	560	5.6K	470K
100	680	10K	1 megohm
150	1K	15K	2.2 megohms

Where They Came From

The numbers used for "standard" values aren't so unusual. The same sequences appear elsewhere, as well—on camera lenses with adjustable f-stops, for example. (The f-stop is a number that indicates how much light the lens will pass through to the film, and is a figure derived by dividing the lens's focal length by its diameter.) While not all the f-stops are marked on every lens, the standard f-stop values are: 1.4, 1.8, 2.8, 3.5, 4, 4.5, 5.6, 6.3, 8, 11, 16, and 22.

Where did these sequences originate? These are *logarithmic* sequences. That is, the numbers are related logarithmically. The logarithm of a number represents the number of times ten must be multiplied by itself to produce that number. Yes, you can do that . . . if you're a mathematician. A characteristic of logarithms is that as they grow larger, the numbers they represent grow larger still. If you double the logarithm, the antilogarithm (the number the logarithm represents) much more than doubles.

Our senses work logarithmically. When we perceive a sound to have doubled in volume, for example, the energy it took to make that louder sound was considerably more than twice as much—three times, in fact. But to make the sound four times louder than it originally was requires

3.3 μF	220 μF
4.7 μF	330 μF
6.8 μF	470 μF
10 μF	680 μF
22 μF	1000 μF
33 μF	2200 μF
47 μF	3300 μF
68 μF	4700 μF
100 μF	6800 μF

Table 4-2. Standard Electrolytic Capacitor Values.

six times the energy or power. In an audio amplifier, a tenfold increase in power results in a perceived increase in volume of a bit over three times. A tenfold increase in volume requires a hundredfold increase in power!

Because we perceive things in a logarithmic way, although we're not aware of it the electronic components that are used to generate the object of our perceptions are also ranked logarithmically. In fact, components such as the potentiometers (variable resistors) used for controlling volume or brightness are available in what's known as "logarithmic taper," which indicates that although when you operate them you think they are working linearly (turning the shaft through twice as many degrees seems to double the volume, for example) they are actually changing value logarithmically.

For this reason, the values of components such as resistors and capacitors, which in the end affect our senses logarithmically, are based on logarithms. It turns out that the "standard" values are actually quite close to the logarithms of numbers whose values are equally spaced.

Variable Resistors

Resistors whose values can be adjusted over a range of values are called by a number of names: *variable resistor, potentiometer* (or *pot*, for short) and *rheostat* are the most common. This type of resistor is marked with its maximum value, and is adjustable over a range from zero ohms to that maximum.

The type of potentiometer usually used as a control (for gain, perhaps, as in a volume control) is made from multiple turns of resistive wire wound on a circular form. A sliding tap that's moved by the potentiometer's control shaft determines how much of the wire the electrical signal being controlled will pass through. The longer the path, the higher the resistance will be, and thus the greater the voltage drop.

Precision multi-turn potentiometers use a material called Cermet (for CERamic METal) to provide resistance. As the shaft of one of these precision devices is turned, it causes the tap to move slowly lengthwise along the Cermet core. The farther the tap moves, the greater the change in resistance. It may take 15 or more turns of the shaft to move the tap from one end to the other, allowing fine adjustment of the resistance. Another factor in making these devices "precision" is the fact that the Cermet core is solid, not a series of wire loops, and the change in resistance is smooth and gradual.

Variable resistors have three terminals for connection to the outside world. One terminal is connected to each end of the resistive material and the third is connected to the tap, or wiper (Fig. 4-1). Depending on

Fig. 4-1. The two end terminals of a potentiometer are connected to the ends of the resistive material wound within the case. The center terminal is connected to a wiper that moves along the resistive windings as the shaft of the device is rotated.

which end of the resistive material you make your connection to, the resistance of the device will increase or decrease as you turn the shaft. You can also determine what effect turning the control shaft clockwise or counterclockwise will have by which of the two resistive-element terminals you use.

In a circuit, the unused terminal (it is extremely rare to make use of both sides of a potentiometer at once; you usually want only an increasing or a decreasing resistance—not both) can be connected either to the center (tap) terminal or to ground, but it should be connected to one of them (Fig. 4-2). It is not a good idea to leave unused terminals "floating"—you want to know what everything in your circuit is doing, even—or, *especially*—if it's not supposed to be doing anything.

Finally, if a variable resistor has three terminals, it's a potentiometer. If it has only two it's a rheostat.

Taper

Potentiometers are available in two *tapers*: linear and logarithmic, sometimes referred to just as "log." The term taper refers to the way

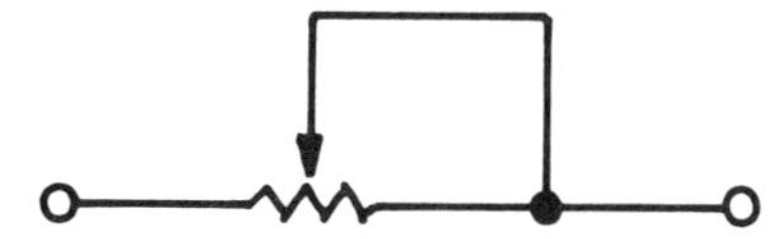

Fig. 4-2. It's good practice always to connect the unused end of a potentiometer's winding back to the wiper.

the resistance changes as the tap is moved, and is a function of the way the resistive material is wound—the spacing between the turns of wire.

Linear taper potentiometers change value evenly throughout their range. That is, turning the shaft of the device a certain amount will always produce the same change in resistance. Turning it through ten degrees, for example, might result in an increase or decrease in resistance of 3000 ohms. Another ten degrees in the same direction would change the value by another 3000 ohms, and so forth.

Potentiometers with a logarithmic taper do not change evenly over their range. They are wound so that the farther you turn them (to increase their resistance, in this example) the faster their resistance increases. For example, the first ten degrees might result in a change of 3000 ohms, but the second ten degrees would produce a change greater than that (perhaps 5000 ohms), and the next ten degrees a change even greater than that. Figure 4-3 shows graphically how the resistance curves of a 100k linear and 100k logarithmic potentiometer might compare.

The reason for having logarithmic-wound potentiometers is that many functions in real life are also logarithmic in nature. As was mentioned earlier, one example of this is sound, which is frequently manipulated (amplified or filtered, for instance) electronically. Our ears and other senses work logarithmically. When we perceive one sound to be twice as loud as another, what we are actually noting is an increase in the energy level of that sound that is much more than double. And at low sound levels, the high- and low-frequency response of the ear drops off logarithmically. (That's the function of the loudness control on an amplifier—to boost the level of the high- and low frequencies so the overall sound appears normal.)

When used in a circuit that controls a signal with a logarithmic function, a log-taper control allows you to adjust the level of that signal in what appears to your senses to be a linear fashion, while actually varying it logarithmically.

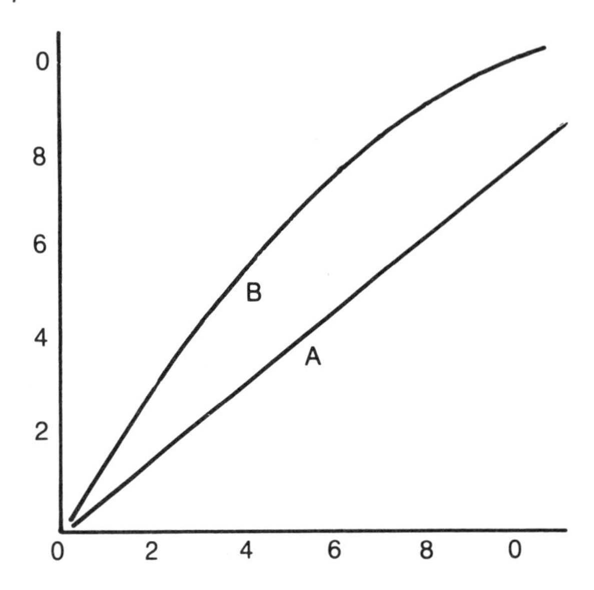

Fig. 4-3. The resistance of a linear-taper potentiometer (*a*) changes at a constant rate over its entire range. In a logarithmic-taper device (*b*), the rate of change starts out slowly and rapidly increases (connecting the device "backwards" will reverse the way the resistance changes to quickly at first and then more and more slowly) as its shaft is turned. This property of log-taper devices makes them well suited to applications that affect our senses, whose sensitivity is also non-linear.

TOLERANCES

Electronics is not always an exact science. For instance, no component has exactly the value marked on it. If you had access to the most accurate test equipment in the world and were to measure the value of any electronic component you picked out of a batch, you would discover that what you had was not exactly what you thought you had. That 5.6k resistor might prove to have a value as low as 5.32k, or as high as 5.88k. And a 10 microfarad capacitor may have a capacitance as low as 8 microfarads, or as high as 18. That's quite a range!

Amazing as it may seem, that sort of range is quite normal in most types of components. It just isn't possible to manufacture devices such as resistors and capacitors in quantity, sell them at an affordable price, and still keep their values within such a narrow range. Of course, you can buy or order components whose values will fall within any range you specify—the degree to which they will match your specifications depends on the price you are willing to pay—but you will generally find that standard-tolerance, off-the-shelf parts will work perfectly well in your applications.

The range of values that can be assumed by a component with a certain marked value is expressed by its tolerance, the percentage by which the actual value can vary from the nominal, marked, value. If a resistor has a nominal value of 100 ohms with a tolerance of 10 percent, its actual value can fall anywhere between 90 ohms (100 – 10% of 100) and 110 ohms (100 + 10% of 100).

To assist you in determining the true values—or at least the range of values they can assume of the components you use, their tolerance is usually marked on them, along with their nominal—intended—values. Sometimes tolerances are indicated by a color code (see below), and sometimes a more complicated letter- or letter-and-number code is used. Resistors usually have their values indicated by a series of color bands (Fig. 4-4), and the last band usually indicates the tolerance. Most resistors have tolerances of 5% or 10%. A silver band on the resistor indicates a 10% tolerance; a gold band one of 5%. Precision resistors, which have tight tolerances of 1% or better, may have those tolerances actually printed on their bodies.

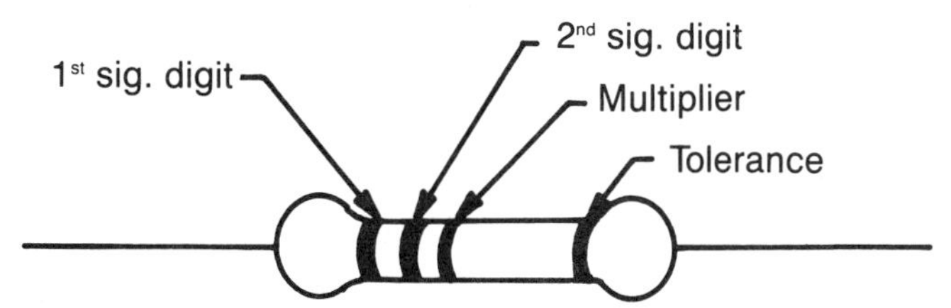

Fig. 4-4. The first two color bands of a resistor indicate the first two significant digits of its value, and the third the number of zeros that must be added to those digits to give the correct order of magnitude. The tolerance band, separate from the others at the other end of the device, is usually gold (5%) or silver (10%), although other tolerances are also available. Most resistors today have tolerances of 5%.

Capacitors, too, are classified according to their tolerances. Table 4-3 shows the most commonly found capacitor tolerances and the letters that are used to represent them. These letters may or may not appear on the devices. The most commonly found category is "M," which represents +80%, –20%. An "M" tolerance capacitor with a nominal value of 0.01 microfarad could have an actual value of from 0.008 microfarad to 0.018 microfarad. As you shall see when we examine how to use capacitors, it is not really important for our purposes where in this range the value falls, as long as it is "in the ballpark."

Capacitors have another characteristic that is important to design

CODE	TOLERANCE
B	±0.1 pF
C	±0.25 pF
D	±0.5 pF
F	±1%
G	±2%
J	±5%
K	±10%
M	±20%
Z	+80%, −20%

Table 4-3. Capacitor Tolerance Codes.

engineers and which is indicated by a letter-and-number code. This characteristic, called a temperature coefficient, reflects the fact that as the temperature of a capacitor changes, frequently so does its value. And, not only does the value change, but it does so at different rates depending on whether the temperature goes up or goes down. This stability with regard to temperature can be very important in circuits where capacitors are used to determine timing constants or frequencies. (Resistors, too, can change value with temperature, but not as radically. Consequently, the question of temperature stability in their case is largely ignored.)

Capacitors that do not change appreciably with temperature are classified as "NP0." Although this stands for "Negative-Positive (. . . temperature coefficient is . . .) Zero," it is pronounced as though the last character were the letter "O," not the number zero. The values of capacitors marked "NP0" are stable, regardless of whether the temperature of their environment rises or falls. You sometimes will find NP0-type capacitors marked with the EIA (Electronic Industries Association) code "COG," which has the same meaning.

The EIA has an established set of specifications for capacitor temperature characteristics, shown in Table 4-4. Thus, a capacitor labeled "Y5P" would exhibit a ±10% variation in capacitance over a temperature range of −30 degrees C. to +85 degrees C.

As with resistors, it is possible to obtain capacitors with almost any tightness of tolerance, but you can pay heavily for the privilege of doing so. Until you get into critical design work where very accurate timing and frequency generation become the most significant things in your life, you will be able to find more important things to worry about than the stability and tolerance of the capacitors you use.

Table 4-4. Capacitor Temperature Coefficients.

LETTER SYMBOL	LOW TEMP. REQUIREMENT	NUMBER SYMBOL	HIGH TEMP. REQUIREMENT	LETTER SYMBOL	MAX. CAPACITANCE CHANGE OVER TEp. rating
Z	+10°C	2	+45°C	A	±1.0%
				B	±1.5%
				C	±2.2%
		4	+65°C	D	±3.3%
				E	±4.7%
Y	-30°C	5	+85°C	F	±7.5%
				P	±10.0%
				R	±15.0%
		6	+105°C	S	±22.0%
X	-55°C	7	+125°C	T	±22%-33%
				U	±22%-56%
				V	±22%-82%

What About Accuracy?

There may come a time when you do need a resistor or a capacitor to be a certain value, perhaps to make a highly accurate oscillator to generate a tone of a specific frequency. There are several ways you can go about this.

Your first thought, perhaps, might be to purchase a few critically accurate components. The question could arise, though, as to *a*) what degree of accuracy you really needed, *b*) how far back in the design of the circuit you wanted (or needed) to carry that accuracy, and *c*) how you were going to afford it.

What really matters in a case like this is the result you get, not how you get it. If you have the super-critical components on hand, well and good. Use them. If not, improvise.

Sometimes you can, believe it or not, make your own precision resistors out of the garden variety kind. The catch is that you need a certain type of resistor to start with, a kind known as carbon-composition. This kind of resistor is not as widely found as it used to be, having been largely replaced by the metal-film type, but you may still come across it from time to time.

You can take one of these resistors with a value somewhat *lower* than the value you require and, with a small triangular file, slowly begin filing a small "valley" in it. The deeper and wider the valley becomes, the higher the resistance will become. If you work slowly and carefully, and check the value of the resistor at frequent intervals as you work, you can hit your target value pretty much on the head. When you've achieved the value you need, you can protect the "wound" you've made in the resistor from degradation by coating it with lacquer or dope of some kind.

"Why," you may be asking, "does the resistance increase if I'm removing some of the resistive material? Why doesn't it go down?" The answer is simple if you stop to think about it. While it's true that you are removing some of the resistor's high-resistance carbon material, look at what you're replacing it with . . . *air*, which, as far as electricity is concerned, has a resistance that's almost infinite. No wonder! Unfortunately, you can't pull the same trick with capacitors.

There is a still easier way to achieve a precision result. Assuming that what you're concerned about is not component values changing with temperature but simply the degree to which the actual value of the components relates to the values marked on them, the "fix" becomes almost ridiculously simple.

First, of course, there's the possibility that all the variations in value among the components will have averaged itself out to zero. That is, some parts may have values a bit higher than marked and some a bit lower, with the net result being equivalent to everything being just as specified. The probability of this happening is pretty low, though. It's probably about as low as that of all the values being high or all the values being low. Therefore, you'd best be prepared to take corrective action.

Since changing the value of any one of the components in the circuit is going to affect the output, why not let the values of most of them wander where they may—within the limits of their tolerances, of course—and use just one of them to correct for the error induced by all the others?

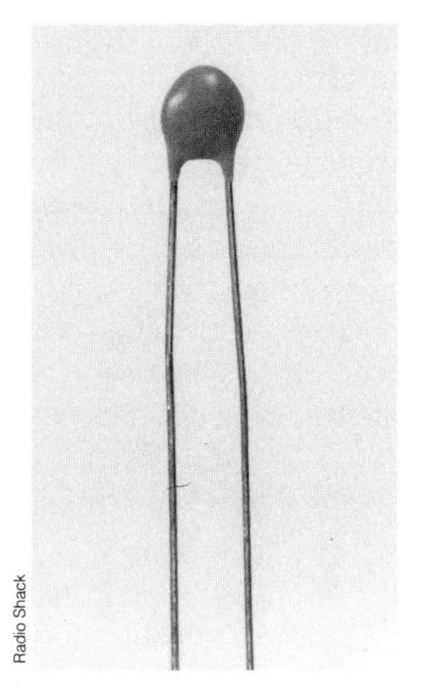

Ceramic capacitors such as this do not usually possess very tight tolerances. They are frequently used for bypassing applications.

In a simple timing circuit of the sort that will be discussed in Chapter 9, the time constant of the circuit is determined by the values of both a resistor and a capacitor. To fine-tune the circuit, though, you do not need precision resistors and capacitors. All that's required is a capacitor that's more-or-less of the value that your calculations have determined is required, along with a *variable* resistor whose maximum value is somewhat greater than the value calculated. You can adjust the value of the resistor while observing the output until you have achieved the correct time constant.

For fine control where large resistances are involved you may want to use a large-value fixed resistor connected in series with a smaller-value variable one. (Remember, the combined value of resistors connected in series is the sum of their individual values.) Thus, to get precisely 1.1 megohms, you might connect a one-megohm resistor in series with a 200k variable one. Assuming that the one-megohm unit was precisely as marked, that would allow you to vary the total resistance from 1 megohm to 1.2 megohms, and would allow a sufficient working range to compensate for any normal variance within the marked tolerance of the fixed unit.

Resistors can also be connected in parallel. Doing so will reduce the combined resistance, since the combined resistance of resistors connected in parallel is less than the value of any one of them. This is an effective after-the-fact technique, since it is usually much easier to connect a resistor in parallel with one already in place than it is to connect one in series with it. Just remember that this method can be used only to reduce combined resistance, not to increase it.

The same sort of thing can be done with capacitors, although the trimmer capacitors you would use are usually very small in capacitance—only a few picofarads—and probably won't have enough range to have much effect. You're much better off making your adjustments in the resistive part of a circuit wherever you can. There are some frequency-determining circuits, such as those involving a capacitor and an inductor (coil) that do not use resistors at all, and where you will have to rely on another type of component to provide the adjustment mechanism.

COMPONENT MARKINGS

If you buy components off the shelf in those blister packs made to hang from pegboard racks, or ask the man behind the counter of the TV-parts place for a 56k resistor, you can be pretty sure of the values you're getting. But what happens when you take the parts out of the bag and they get all mixed up with the others in your junk box? Or, what if you just got a handful of components out of someone else's junk box to begin with? How do you know what you have, and how do you find what you're looking for?

That's why manufacturers mark components with their values and other information you may need. As you'll quickly find out, though, the markings are not always a straightforward indicator of what you have in hand. They take some interpretation.

Color Codes

Probably because it was originally impractical to print the numerical values of resistors and capacitors on their bodies, these values were indicated by a series of color bands or dots painted on them. While, thankfully, this practice has disappeared as far as most capacitors go, it is still used on resistors.

Typically, resistors are marked with three color bands to indicate their value; a fourth band of silver or gold (see above) shows the tolerance (Fig. 4-4).

Let's concentrate on the three primary bands. The first two of them, starting from the left end (keep the gold or silver tolerance band toward the right—it's easy to do this since neither gold nor silver is used in the three primary bands) represent a two-digit number. The third band is called the multiplier, and represents the number of zeros that have to be added to that number to give the value of the resistor.

Here's a way to associate the colors with the numbers they represent. ***Black***, which is really no color at all, stands for zero. It's a more important number than it appears. ***Brown***, which you might think of as a sort of light black (or maybe the first color you think you see as the sun rises) is 1. At the other end are *gray* (8) and *white* (9), as the light of day gets brighter and brighter. We don't need a color for ten, that would be a brown band and a black one.

The numbers between—two through seven—are represented by the colors of the spectrum (or rainbow). This spectral sequence used to be taught as VIBGYOR (pronounced ''vib-gy-or''), and stands for: Violet, Indigo, Blue, Green, Yellow, Orange, Red. Disregarding indigo (a sort of purple), the electronic color code uses violet for 7, blue for 6, green for 5, yellow for 4, orange for 3, and red for 2. Of course, this sequence is backwards compared to the letters in VIBGYOR, but once you've made the associations, that won't matter too much. Besides, VIBGYOR is easier to pronounce than ROYGBIV. A little, anyway.

The third digit of the three-digit number, called the *multiplier*, tells you how many zeros to tack on to the other two to indicate the magnitude of the resistance. Since resistance values run in cycles as we've seen, a resistor whose first two digits are five (green) and six (blue) could have a value of 56 ohms, 560 ohms, or maybe 56k. The number of zeros indicated by the third color band makes this clear.

The same colors are used as for the first two bands. A black band indicates that no zeros are to be added. The value of the resistor is the value of the first two digits alone. The green-blue resistor with a third, black, band is 56 ohms. A brown band tells you to add one zero (560 ohms in our case), a red band two zeros (5600 ohms or 5.6k), and so on.

You will probably never see a multiplier band higher in value than 6 (blue) since resistor values higher than 20 megohms (20 followed by six zeros) are quite rare. If you do see what appears to be a high-order multiplier, double check to make sure it doesn't represent something else as may be the case with some precision resistors.

Sometimes resistor manufacturers hand out small paper or plastic devices with rotatable wheels within windows that allow you to translate the color bands into values. These can be fun for a little while, but what will you do when you lose yours? If, instead, you force yourself to do the colors-to-digits translation for a little while from the beginning, it will soon become second nature and you won't have to rely on mechanical devices to do the job for you.

A few capacitors, especially the older high-voltage mica types, used a series of color dots to indicate the value of the device. This is shown in Fig. 4-5. With any luck, you'll never run into this system.

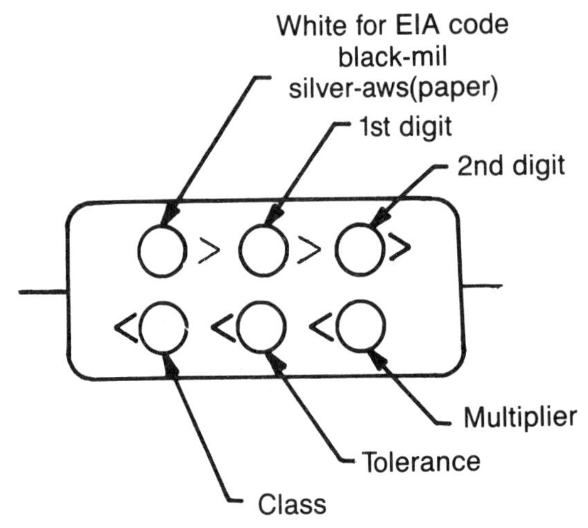

Fig. 4-5. Although most capacitors have their values more-or-less clearly printed on them (see text) you may still come across some small high-voltage types that use colored dots. The colors represent the same digits as do those used to code resistors.

Sometimes the value of a component is printed right on it. This is often the case with precision resistors and with some capacitors. Tubular and can-style electrolytic capacitors are the simplest to decipher; their values are usually printed in ordinary numbers and letters that anyone can understand. It would be no problem to identify a capacitor marked "35V, 2200μF." Sometimes the voltage is identified as "WV" (Working Voltage), and sometimes the letters "DC" (direct current) are added.

Capacitors are frequently so small, as in the case of dipped tantalums (or worse, surface-mount leadless chip capacitors) that there isn't space on them to indicate their specifications in full. A system that combines abbreviation with positional notation is used. It can work in several ways.

The first way uses numbers and letters. A typical marking might be "4R7" The letter "R" serves to mark the position of the decimal point in the value—that's all it does, but it allows us to know that the value of the capacitor in question is 4.7 *somethings*, the units (microfarads or picofarads) generally being apparent to you, the experienced user.

Another system uses a three-digit number. The first two digits represent the first two digits of the value, and the third is a multiplier indicating the number of zeros that have to be added to the base number to make it read as intended. Depending on the size and type of capacitor, the multiplier can express picofarads or microfarads. Electrolytics usually have values in microfarads, most other types have capacitances expressed in fractions of microfarads or in picofarads. A small film capacitor might have a value shown as "104." This represents 100,000 pF. (10 followed by four zeros, as indicated by the multiplier). If you divide by a million (remember, a picofarad is a micro-microfarad—a millionth of a millionth of a farad) you get the value in microfarads, 0.001μF in this case. (The easiest way to make the conversion is to subtract six zeros. If there aren't enough zeros in the original figure to make six, you can add the "missing" zeros to the right of the decimal point *before* the significant digits.)

ELECTRICAL CHARACTERISTICS

So far we have discussed the electronic characteristics of components—resistance and capacitance. There are other factors to consider as well, their electrical characteristics. For us, this means working voltages and power-handling capacity, or wattage.

Working Voltage

Capacitors are often marked with their working voltage. This is frequently indicated as WV or as WVDC (the DC stands for direct current). The working voltage of a capacitor is very important, since if you ignore it you may find that the device will self-destruct shortly after you put it into service.

Capacitors are pretty fickle devices. One of the things they don't like is more voltage than they're built for. This can result in internal arcing—sparks that punch holes through the layers of metal foil forming the electrodes of the device and render them useless. (Some types of electrolytic capacitors are self-healing and can recover from such wounds, but it's better not to tempt fate.) To prevent such damage, the working voltage—the voltage beyond which damage may occur—is given.

If you are working from a schematic published in a book or magazine you may find the working voltage of the capacitors indicated in a small note off to one side of the drawing that reads something like, "All capacitors 35 WVDC unless otherwise specified." A capacitor, especially one of the electrolytic types found principally in power supplies, may have its

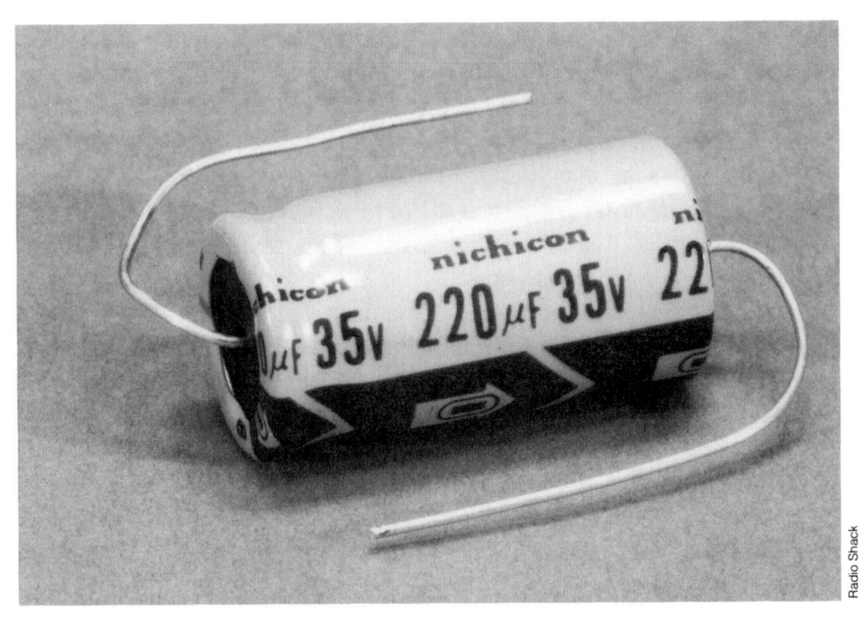

Tubular electrolytic capacitors come in *axial* and *radial* versions. This one, with its leads extending from both ends along its axis, is an axial type. Radial types have both leads protruding from the same end, along a radius.

value indicated as 100/35V. This indicates a capacitor with a value of 100μF and a working voltage of 35 volts. Occasionally, you have to use your own judgment in the matter of working voltages.

It's a good idea to over-engineer your circuits in the area of working voltages. It would not be out of line for you to use components with working voltages 50% or more greater than those you expect them to encounter. For instance, if the highest voltage you anticipate in a circuit is five volts, it wouldn't hurt to use capacitors with working voltages of eight or ten volts. You never know what you're going to run into, nor do you know how close a manufacturer is pushing his tolerances. The 50% margin will protect you against most voltage surges, and will, in general, ensure a longer and more stable component life.

Sometimes you cannot find exactly the component you want. Perhaps your calculations call for a 0.01μF capacitor with a working voltage of 16 volts. What you have on hand, though, or what you can find, is marked 50 WVDC. Can you use that part? Of course you can! The working voltage of the component is not going to affect its electronic performance; it is more a measure of its electrical characteristics. The capacitor will probably be a little larger than one of the same value but a

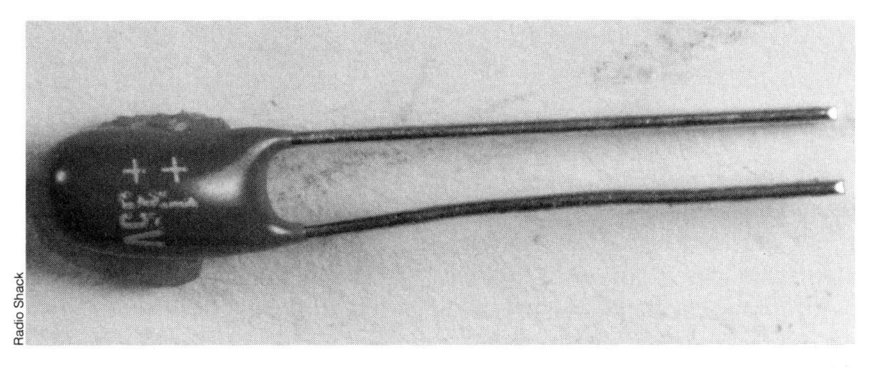

Radio Shack

Dipped-tantalum electrolytics have fairly tight tolerances and are very stable. They are also rather expensive and their use can be reserved only for situations where their characteristics are specifically required.

smaller working voltage, but if you can squeeze it in, there's absolutely no reason why you shouldn't use it.

Wattage

When an incandescent light bulb glows it does so because an electric current is passing through the filament of high-resistance material. Enough heat is lost to friction in fighting that resistance to cause the material to glow. The same thing can happen in the resistors used in an electronic circuit if they are not chosen properly.

If you try to squeeze too much current through a resistor it will first get very warm, then hot, and eventually will burn itself up—along, possibly, with portions of the circuit surrounding it. The wattage, or power-handling capacity, of a resistor has to be matched to the circumstances of voltage and current in which it will be used.

Fortunately, the low voltages and signal levels required by the semiconductor devices used in most signal-generating and processing circuits ensure that the low-wattage devices commonly available are more than adequate to do the job. For instance, resistors are most readily available today in $^{1}/_{4}$-watt versions. Considering that voltages will rarely exceed 12 volts and that currents will be on the order of only a few milliamperes, the formula for power ($P = E \times I$) indicates that a $^{1}/_{4}$-watt (0.25-watt) resistor could handle about 20 mA. In a five-volt circuit, this resistor could handle 50 mA. This, believe it or not, is enough to drive a small loudspeaker or pair of headphones quite loudly. Where signals are simply moving from one stage of a circuit to another, 50 mA is huge! In some circuits you can get away easily with $^{1}/_{8}$-watt resistors. (If you can find them. They are becoming more available in surface-mount varieties, but these are difficult to work with.)

If you have a question about how hefty a resistor you need in a particular spot, it is easy enough to find out. Again, use the equation for power. If you know the voltage and current that the resistor will be handling, you can calculate the wattage involved with ease. Make sure that the power-handling capacity of the resistor is at least equal to the figure at which you arrive. As with working voltages in the case of capacitors, extra power-handling capacity will not hurt, and you should never push components to their limits.

Resistors are readily available—if not in a local store, then by mail order, in 1/4- and 1/2-watt sizes (Fig. 4-6), and in 1-watt and larger sizes as well. Large (10- or 25-watt) wire-wound "power" resistors, usually used in power supplies, are frequently available at TV-repair shops or parts-supply stores. In a pinch, you can make your own high-wattage resistors by connecting resistors in parallel. Just remember that you must allow for the effective resistance of that part of the circuit. The easiest way is to use two equal-value resistors. Each should have a resistance twice that desired. As an example, two 1/4-watt ten-ohm resistors in parallel will give you what is effectively a 1/2-watt five-ohm one.

You can also connect capacitors in series to increase their working voltage, but remember that the effective capacitance of the string will be reduced.

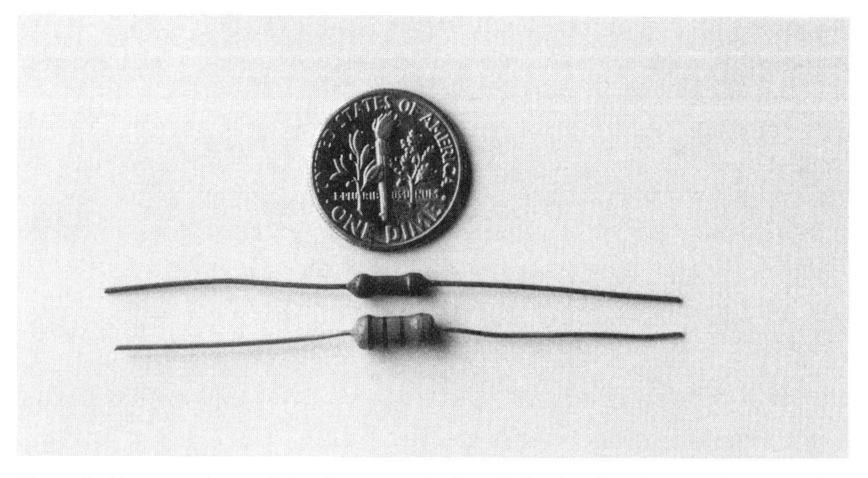

Fig. 4-6. One-quarter-watt (top) and one-half-watt (bottom) resistors. Even smaller one-eighth-watt devices are available, but they are harder to locate.

Heat Dissipation

Associated with the power-handling capacity of a device is its ability to dissipate the heat generated. This too is given in watts, and is an important factor in selecting semiconductor devices for use in circuits such as power supplies and amplifiers where high currents are encountered. We'll discuss this subject in the next chapter, and further in Chapter 8, which deals with power supplies.

5
Discrete Semiconductors

The components we've considered so far—resistors and capacitors—are generally considered to be secondary in importance to the "glamour" devices such as transistors and integrated circuits. These are the ones that do things! In this chapter we'll examine the two most basic semiconductor devices, diodes and transistors.

DIODES

Diodes are the most elementary of solid-state devices. And besides being useful in and of themselves, they are also the foundation of every other semiconductor component, be it a simple transistor or the VLSI integrated circuit at the heart of a microcomputer.

A diode is made of a piece of N-type semiconductor material having a surplus of electrons joined to a piece of P-type material, which has a deficiency of electrons. The N-type material end of the device is known as the *cathode* and the other end as the *anode*. These names are hold-overs from electron-tube days.

The schematic symbol for a diode is shown in Fig. 5-1. The flow of electrons in a diode (or other junction-type semiconductor) is against the direction of the arrow. The cathode end of a semiconductor diode is marked with a printed band around the device. Diodes use the prefix "1N" in their part numbers; any component whose part number begins with "1N" is a diode.

Diodes are usually manufactured using one of two elements, silicon or germanium. Silicon diodes are generally used in power-supply appli-

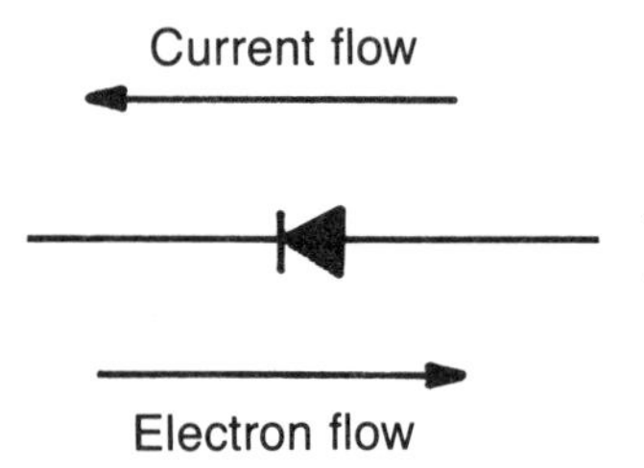

Fig. 5-1. In diodes (and other semiconductors) traditional current flow is always in the direction of the arrowhead. Electron flow, however, is always in the opposite direction—*into* the arrow.

cations while germanium types, such as the 1N914 and 1N4148 which are more sensitive and respond more quickly to change, are found in RF applications and in switching situations where high speed is essential.

Diodes have associated with them a phenomenon known as *voltage drop*. That is, when a voltage is applied across a diode, a difference in potential can be measured across it (Fig. 5-2). In the case of silicon diodes that voltage drop is about 0.7 volt, and in the case of the more

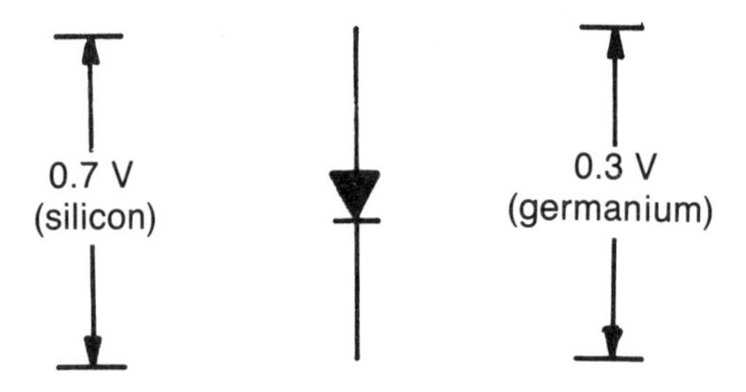

Fig. 5-2. Silicon-based semiconductors exhibit a voltage drop of 0.7 volt across them. Germanium devices, which may be used for switching or high-frequency applications, have a voltage drop of only 0.3 volt.

Electron Flow/Current Flow

While electrons travel "into" the arrow in the schematic symbol of a diode or other semiconductor, current flow is said to be the other way, in the direction of the arrow. It all gets very confusing, and is due to the assumption made in the very early days of electricity that an electric current flowed from the positive pole of a device to the negative one—nobody back then knew anything about the existence of electrons. We're stuck with this convention, though. Try to remember that the electron flow is in the direction opposite to the traditional current flow.

sensitive germanium type it is 0.3 volt. This voltage drop is a very important fact of life in the design of circuits using diodes, or in circuits using components such as transistors and integrated circuits. The concept of bias, as discussed on page 73, is an illustration of how important this voltage drop is. In this book we'll be concerned with silicon diodes.

The most important property of diodes is that they permit electrons to flow in one direction, from cathode to anode, but not in the other. Because of their "one-way" properties, diodes are used in power supplies (see Chapter 8) to convert alternating current to direct current. In this capacity they are called "rectifiers," from the Latin word meaning "straight" or, in this case, "to straighten." The most common variety of diode is the "1N4000" series, silicon devices that can handle an ampere of current. Table 5-1 lists the most readily available of the

Table 5-1. Rectifier Diode Family.

1N4001	50V
1N4002	100V
1N4003	200V
1N4004	400V
1N4005	600V

1N4000-series diodes, showing the voltage rating of each. It's a good idea to allow for 50 percent higher voltage than you anticipate having to handle, and in the case of diode rectifiers, the cost of this extra protection is negligible.

The characteristic voltage drop across diodes leads to several interesting applications for them. One of these is to provide a sort of false ground in power supply circuits. By connecting a power supply component such as a three-terminal regulator to ground through a diode, or string of diodes, it's possible to change the output voltage of that device to one other than that for which it was designed. This will be explored in Chapter 8.

Because of their "one-way" properties, ordinary silicon diodes are often used as protective devices, to prevent current from flowing in portions of a circuit where it is not wanted.

There is a special type of diode called a *Zener* (Fig. 5-3) that is also used in voltage regulation applications. This type of diode has associated with it what is known as a Zener voltage, analogous to the 0.7-volt drop across an ordinary silicon diode. That is, the voltage across a Zener diode must equal its Zener voltage before it will begin to conduct. This makes Zener diodes useful in power-supply applications where they can be used as regulators or where they can be used in place of a string of silicon diodes to place a device on a ''pedestal'' above ground. Zener diodes are connected with their cathode ends *away* from ground.

Fig. 5-3. A Zener diode, useful in establishing a voltage reference in a circuit, can be manufactured to have almost any voltage drop desired across it.

Zener diodes are available with ratings ranging from about three volts to several hundred volts, and with power-handling capacities from a few dozen milliamperes to several amperes. If you come across a device that looks like a diode, but has a strange part number—1/4 M2.4A210, for example—it may be a Zener.

WHY THERE ARE SO MANY TRANSISTORS

Selecting the one or two transistors your circuit might require from among the thousands and thousands that are readily available (and the thousands more that you discover as soon as you really start looking) can seem a formidable task. It isn't, really, once you know what's behind that proliferation.

Maybe the parts list for the project you're assembling calls for ''any general-purpose transistor.'' So you leaf to the back of your favorite electronics magazine, or pull out the pile of catalogs you've accumulated as the result of filling out reader-service cards month after month, and look to see who's got a good deal on general-purpose devices.

Let's see, where are they? ''RF transistors?'' ''Power transistors?'' ''Small signal transistors?'' ''Switching transistors?'' Since you're building a little audio amplifier, you probably want ''small signal'' transistors; after all it'll be handling only a small signal. Let's see what there is in that category: Hmmm . . . 2N2222, 2N2222A, 2N3904, 2N3906, and about fifteen other types. Maybe your transistor data book will give you a better idea.

Well, that's no help either. With something on the order of 30,000 different types of bipolar transistors alone, there are at least fifteen pages of small-signal devices (Fig. 5-4), and only one or two of those also appear in your mail-order catalogs. How are you going to decide which is the right one? Why are there so many transistors, anyway?

Selector Guides 1

The following selector guides highlight semiconductors that are the most popular and have a history of high usage for the most applications.

These selector guides cover a wide range of small signal plastic and metal can semiconductors.

A large selection of encapsulated plastic transistors, FETs and diodes are available for surface mount and insertion assembly technology. Plastic packages include TO-226AA, TO-226AE 1 Watt and SOT-23. Plastic multiples are available in 14-pin and 16-pin dual-in-line packages for insertion applications: SO-8, SO-14 and SO-16 for surface mount applications.

Metal can and ceramic packages are available for applications requiring higher power dissipation or having hermetic requirements. TO-18, TO-205AD, TO-46, TO-52 and TO-72 packages contain discrete devices. There is a variety of ceramic dip and flatpacks available for multiple transistors, FETs and diodes.

Devices which are JAN, JANTX, JTXV or CECC qualified are noted in the individual selector guides or in the Hi-Rel and Military Section of this selector guide.

Table of Contents

Fig. 5-4. This contents page from the Motorola *Small Signal Semiconductors* data book gives you some idea of the range of devices that are available in this category alone.

Transistor Classes

Even the most confused parts hunter will have to agree that it makes sense to divide transistors into classes, according to the purpose for which they will be used.

For example, it would be foolish to use a bulky heavy-duty power transistor that cost several dollars to do a job that could be handled just as well by a considerably smaller and less expensive device that was not overengineered for the job. The question is, which specifications made it "just right?"

Each class of transistor is designed specifically for a particular class of operation, and it is the characteristics that make it suitable for a particular type of job that set it apart from the other classes.

In some cases the main criterion is the highest frequency at which a device will provide its rated performance. This is frequently expressed at f_T, the frequency at which gain drops to unity. RF transistors, for instance, can have f_T's of hundreds or even thousands of megahertz, while those good old "general purpose" types can get by with a considerably lower high-frequency limit.

Gain, shown in the data books as h_{FE}, is another factor used to divide devices into classes. It may be a matter of design, or simply one of practical limitations, that determines the gain of a particular transistor. Power transistors, for example, are built primarily to handle large loads. While their gain may not be as impressive as that of other types, that is not important. What matters is that they can handle high currents or voltages without overheating, melting, exploding or otherwise rendering themselves inoperable.

Another important characteristic is breakdown voltage. That is, how high a voltage can a particular device handle? Is it millivolts, volts, or kilovolts? This factor can influence the size, packaging, and operating characteristics of a transistor, as well as its price. There are small-signal transistors designed to handle only a relatively few volts and milliamperes of current, and there are power transistors intended for massive amperages. Depending on their technology (bipolar, FET, MOSFET, etc.), application, and packaging, transistors can be divided into more than three dozen classifications.

Where Transistors Come From

None of those devices in the lengthy list in the data book slipped in while no one was looking. They all got there for a reason.

Many new designs come about as the result of advances in technology, either new developments or improvements of existing manufacturing techniques. When it became practical to produce power MOSFETs in commercial quantities, a whole new product line was launched. This did not mean, of course, that all the types of power transistors already in existence were made obsolete and could be dropped from the catalog. Rather, these older products continued to be manufactured because there were many circuits in which they performed perfectly well, and in which they could continue to be used.

Quite a few new products arise from the design engineers' requirements. The calculations for a particular circuit may call for product specifications that cannot be met, at least within the specified tolerances, by any existing device. If a large enough quantity of a new-specification device is required, or if the manufacturer approached about this new product sees a general-market potential for it, he will set his engineers to work, and the product will eventually be manufactured and wind up in the next data book or supplement.

A small manufacturing company in search of a transistor will take the time to investigate devices that are already available in the hope of finding one that will do the job it has in mind. Sometimes it will even pay to redesign a circuit around an existing device to take advantage of its characteristics, or to avoid the expense of having a special part designed and manufactured.

It can be more economical for a large company, or one that anticipates the purchase of millions of a particular device, to request that a component be produced to its precise specifications. It is cheaper for the company to take that route than it is to use a device that perhaps already exists but whose future supply cannot be guaranteed. This specially requested part may be identical in all but one minor characteristic to another that already exists, but no semiconductor manufacturer is going to turn down an order worth millions of dollars. He'll produce that device, and another part number will find its way into the data books.

Adding to the proliferation of device types is the fact that the manufacture of semiconductors is not an exact process. Certainly, it is much less of an arcane art, and much more of a science, than it was many years ago. There are still a number of variables in the process, however, not all of which can be controlled exactly. For this reason, during a manufacturing run there are always a certain number of devices produced which either fall short of, or perhaps exceed, the tolerances required in one or more areas. The areas of gain, breakdown voltage, and power dissipation are among those subject to variance in the manufacturing process.

The devices that exceed the specifications can be marketed as a premium product for those who need their superior characteristics. The cutoff frequency of a part, for instance, might be specified as 200 MHz but in achieving that mean, a number of devices might be produced with a much higher cutoff frequency, say 300 MHz. While these could be used by the original specifier, they are really too good for his purposes. However, there are probably other places in which such a high-performance part could be used, so it is culled from the run and made available on its own. At a premium price, of course.

Some devices meeting particularly high tolerance requirements may be classed as ''mil-spec'' or ''high-rel(iability)'' parts. These must meet stringent testing criteria, and if they pass they will get their own part numbers.

Similarly, there are a significant number of components produced whose cutoff frequency is only 125 MHz. The engineer who ordered them certainly cannot use these parts in his design, but that is not to say that they are worthless and have to be discarded. Somewhere out there is someone whose design needs are not so critical, and who can make good use of these out-of-spec components. Indeed, many ''general purpose'' components enter the market in just this way.

It is possible, within the limits of statistics, to predict the number of devices that will not meet specifications, and there are frequently enough of these—perfectly usable parts, except that they are not exactly what was ordered—to be included in a catalog as a new product with a part number of its own.

This is not to say that the quality control techniques, or manufacturing procedures used by one or another manufacturer are at fault. It's simply the way the game is played—some parts meet the specifications and some don't.

The Case for Cases

It is not only the ratings and electrical characteristics of a particular device that set it apart from all the others. The same part may be packaged in several different ways, and each new package gets its own part number, or at least a suffix that differentiates it from the others (Fig. 5-5).

MD7021,F
MQ7021

MD7021
CASE 654-07, STYLE 5

MD7021F
CASE 610A-04, STYLE 1

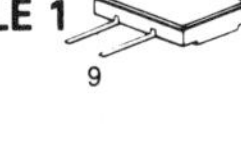

MQ7021
CASE 607-04, STYLE 1

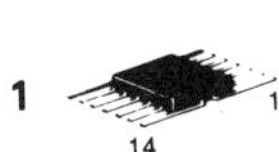

COMPLEMENTARY
GENERAL PURPOSE TRANSISTOR

NPN/PNP SILICON

Fig. 5-5. The same device can appear in a number of different cases, each time with a different designation. The packaging may affect some of the device's operating characteristics.

Certain applications require "ruggedized" parts that can withstand extremes of temperature or humidity, or simply rough physical treatment, and for these a special packaging is required, possibly ceramic or metal rather than plastic. Alternatively, a part that was originally specified to withstand extremes of environment might be worthwhile producing in a less rugged version for more ordinary applications. Bingo, another part number.

Power dissipation is another factor that can add part numbers to the data books. The power-handling capacity of a semiconductor device can vary according to the way it is packaged. In a plastic TO-92 package a transistor will have a power-dissipation factor of only a few hundred milliwatts. In a TO-220 tab-type package that rating can increase to a few watts, and in a heavy duty metal TO-3 package it will be even higher. What's inside each package is the same; the package characteristics change the characteristics of the device. Sometimes the part numbers are the same, with only the suffix changed to denote the difference in packaging but sometimes a new package gets an entirely new number. Some of the common transistor packages are illustrated in Fig. 5-6.

Size and shape are also important. Someone designing a high-density circuit board may require that a component be packaged in a case of a particular shape or size, so he can fit as many devices as possible in a given space. Such custom packaging is not cheap and a manufacturer will deem it worthwhile to make the same part available in a cheaper, more conventional, package.

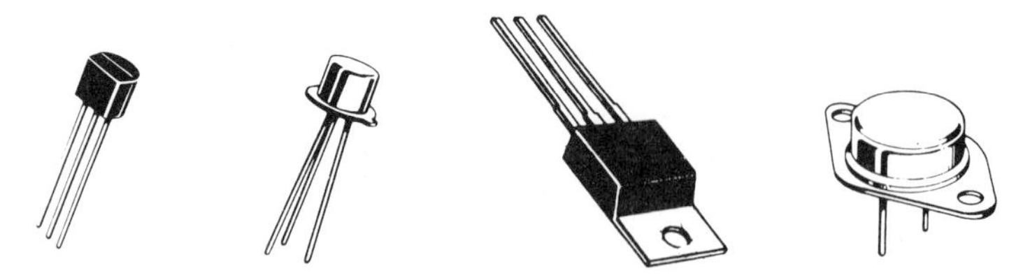

Fig. 5-6. These are the semiconductor packages you'll most frequently encounter in your work. From left to right they are: TO-92, TO-18, TO-220, and TO-3. The letters "TO" stand for "Transistor Outline."

Where Do They Get Them?

As you browse through the catalog from your favorite supplier searching for parts, it's obvious that the selection of transistors he offers is considerably smaller than the range of offerings from even the smallest manufacturer. How do these people narrow down the choices?

The line stocked by these vendors depends principally on two factors: availability, and customer need. What this boils down to is the familiar law of supply and demand.

Manufacturers—both of transistors and of the devices that use them—frequently find themselves with more of a particular part than they need. There is nothing wrong with the quality of that part, it's just that for one reason or another they have too much inventory. In such cases, they reduce that inventory by selling off the excess. There are firms that specialize in disposing of this excess inventory, taking it off the overstocked company's hands for a price that's probably less than was originally paid or asked for it. Sometimes, too, the over-inventoried manufacturer can sell directly to a vendor through an ad indicating his excess stock placed in the "Classifieds" section of a trade newspaper. The low price you pay is not a reflection of the quality of the item, but rather of the sacrifice the overstocked manufacturer had to make to dispose of his excess.

Many of the devices in a vendor's catalog are there simply because they are consistent best-sellers. As long as people keep buying 2N2222s, these transistors will continue to appear in vendor's price lists. Depending on how plentiful or scarce they are their price will vary, but they will continue to be there.

Sometimes, too, a device shows up in a catalog because it is frequently "special-requested." Vendors receive many phone calls and letters from people trying to locate such-and-such a part. Such requests sometimes arise in response to specific projects that appear in electronics magazines.

It might be that the person responsible for the project used an "orphan" transistor he found at the bottom of his junk box. Some people, for one reason or another, are unable to find a more readily available substitute for that part and so request the oddball one. If there are enough such requests, a vendor will see a profit in that item, and add it to his list. It should be noted that sometimes, albeit rarely, there is no readily available substitute for a particular component; this too is a reason for it to be stocked. If there are enough requests for an item, a source for it will be located, and it will find its way into the vendors' catalogs.

How Do You Choose?

It's a lot easier to select the part you need for your project than it might at first appear from the plethora of transistors that present themselves. All it takes is some common sense.

The first thing to do to narrow down the possibilities is simply to determine into what category the part falls. If you're building a huge audio amplifier, you need a power-type device. If your project is going to operate at moderate voltage and current levels—perhaps in a more modest audio device of some sort— a small-signal transistor will do the job nicely. If what you are building is going to oscillate or switch at high frequencies, a switching transistor is called for, although at audio frequencies that same small signal-device—which is probably what is intended when a "general-purpose" device is specified—will do as well.

In most cases you don't even have to look at the specs. Unless you're "pushing the edge of the envelope"—working somewhere out at the outer fringes of a category—almost any transistor in that class will do the job.

If you want to be more particular, you can start scrutinizing the specifications. The thing to bear in mind as you read them is that your application and the specifications do not have to match perfectly everywhere. If your amplifier is going to have to handle frequencies up to 100 kHz, there's no reason why the cutoff frequency of its components has to be exactly that (or a little bit more, if you're going to follow good engineering practice and overdesign a bit). A device that's usable all the way up to one MHz will work just as well at one-tenth that frequency as it will at the high end of its specified range. As long as your application falls within those extremes, you can use the device. And if you get the "higher-quality" part at a bargain price, so much the better.

There are several computer programs on the market intended to simplify the process of the design of electronic circuits. One of them has a library of about 12,000 transistor devices from which to choose. Another offers a library of only 200 or so. The rationale behind the smaller selection in the second program is that, as we have discovered, there really aren't that many different transistors—it just appears that way. Many parts are just the same device in a different costume, or a device that behaves slightly differently from its neighbors when pushed to the limit.

If you stick to the middle ground, though, it will be hard for you to go wrong. Once you're in the ballpark, as it were, there is no absolutely right or wrong device.

NPN OR PNP?

The first question you encounter when choosing a transistor is: "NPN or PNP?" The letters refer to the way the transistors are constructed. Devices that are designated "NPN" are made of a "sandwich" of a layer of positively doped silicon (which has a deficiency of electrons) between layers of negatively-doped silicon having a surplus of electrons. PNP transistors consist of a negatively-doped layer between two positive ones.

The use of one type or the other is generally dictated by the environment in which it will be applied. While the specifications of NPN types seem usually to be a little better than those of complementary PNP types with perhaps slightly better gain and frequency response figures, until you get into super-critical circuit design—and devices such as little audio-frequency amplifiers do not fall into that class—these differences will not really matter to you.

What will matter is the polarity of the voltages in the circuit in which the device will be used. What is the polarity of the voltage on the collector? If it is positive you will need an NPN device; a negative voltage will require a PNP one.

Finally, transistors are frequently used a switches. Applying a voltage to the base of the device turns it on; removing it turns it off. The polarity of this control voltage can determine the type of device that will be required in a particular situation.

The Importance of Being Biased

One of the most important concepts associated with the operation of semiconductor devices such as diodes and transistors is that of *bias*. If you understand what bias is and how to apply it to your needs, you can make almost anything work. Sometimes bias seems to be what electronics is all about.

The word "bias" means "slant" or "lean" (in a particular direction). When, for example, someone holds a biased opinion, that opinion "leans" in a certain direction, perhaps in favor of or against the guilt of someone on trial. And, just as opinions may be biased and so cause decisions (of a jury, perhaps) to go one way or another, so can semiconductors.

The N-P or P-N junction in a silicon diode or transistor has associated with it a voltage drop of about 0.7 volt. If a voltage of less than 0.7 volt is applied across the diode, no current will flow. That 0.7-volt difference in potential has to be overcome before things start happening! If a voltage greater than 0.7 volt is applied, it will drop by 0.7 volt.

To get a current flowing through a semiconductor it is necessary to bias it in the direction of current flow—that is, make it lean in that direction by giving it a "push" of 0.7 volt. This is called *forward bias*.

Bias voltage can also be added to a circuit or part of a circuit to raise it to a certain potential above ground. This is done in amplifiers to establish their characteristics—whether they will pass all of an ac waveform, as in a Class A amplifier, or whether they will pass just a portion of it, as in a Class B or Class C device. Sometimes *reverse bias* is applied to a device to prevent current from flowing, either entirely or below a certain point.

BIPOLAR TRANSISTORS VS. FETS

Field-effect transistors, more commonly known as FETs, are at the high end of the transistor spectrum. They are generally faster, cleaner, and in general closer to the "ideal" transistor than their bipolar cousins. They are also more delicate, being more sensitive to being "zapped" by static electricity discharges. Unless a circuit specifically requires it, an ordinary bipolar transistor will do for most experimental work.

PNP and NPN

The way a transistor is constructed is shown in simplified form in Fig.5-7. The drawing on the right shows an NPN device—one constructed of a layer of positively-doped silicon (with a deficiency of electrons) sandwiched between two layers of negatively-doped silicon (with a surplus of electrons). The drawing on the left shows a PNP device, which is put together the opposite way. The names of the layers, and the way the leads are attached and labeled, are clearly indicated.

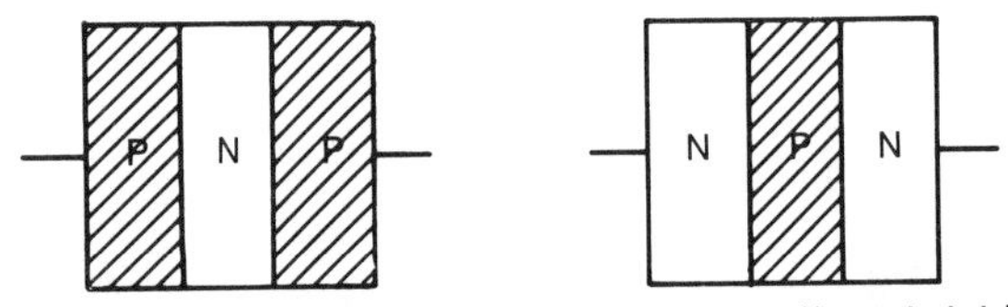

Fig. 5-7. A simple bipolar transistor can consist of a "sandwich" consisting of N(egative)-type material having a surplus of electrons between two pieces of P(ositive)-type material with a deficiency of electrons as shown at the left, or the opposite, shown at the right. The former type of device is known as "PNP" and the latter as "NPN."

Figure 5-8 shows the schematic representations of these same transistors. The arrow on the emitter not only indicates that that is the emitter, but also shows which way electrons (and therefore current) flow in the device. As in a diode, the current flows in the direction of the arrow, which means that the electrons are actually moving in the opposite direction.

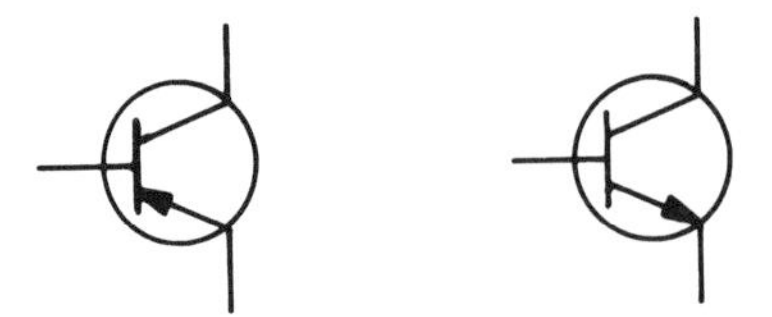

Fig. 5-8. If you can't remember which transistor symbol represents which type of device, try this: Think of the emitter (the element with the arrowhead) as a compass needle and the base (the large "slab") as the North Pole. "PNP" (left) then stands for "Point North Please," while "NPN" (right) refers to "Not Pointing North."

6
Integrated Circuits: Digital and Analog Devices

Integrated circuits—also known as ICs or, sometimes just "chips" (from the tiny chip of silicon on which the circuit elements are built up), are what their name implies: entire electronic circuits integrated on a single piece of silicon. They are composed of microscopic transistors, resistors and capacitors—sometimes hundreds of thousands of them—all on a little silicon substrate about the size of your fingernail.

ICs have turned the design of electronic circuits into something akin to ordering dinner from a menu in a restaurant, as opposed to cooking it for yourself. You don't have to prepare anything from scratch; you just point at what you want and you get it. When you buy an IC, you buy it by its function. You don't buy five transistors, four resistors, and two capacitors; you buy an audio amplifier, or a two-input NAND gate, or a five-volt voltage regulator. This can make most of the problems of circuit design almost trivial. However, to the newcomer, the variety and complexity of the types of ICs available can be bewildering.

Broadly speaking, integrated circuits can be divided into two types: digital and analog. Those in the first category, also known as logic ICs, are the type generally associated with computers, although they have many other uses as well. They live in a special world: their inputs and outputs are usually restricted to just on and off states. These are represented by the presence of a voltage or the absence of one, typically represented by five volts and ground. It's incredible, but whole computers can be built and function around just two states: on and off, yes and no, five-volts and ground.

Logic ICs are also used in computer memory (RAM), switching applications, clocks, and in other areas where operations can be performed by, or signals translated into, a form that can be represented by just two states.

Analog ICs are those that deal with and process signals that are not restricted to a couple of discrete states, but that vary continuously, much as things do in the real world. Amplifiers, voltage regulators, the signal processing circuits used in TV sets—all are analog ICs. They will be discussed later in this chapter.

Although most of this book concerns itself with analog circuits, a brief digression into the world of logic ICs will help prepare you for life in an increasingly digital world.

WHAT IS LOGIC?

Logic circuits are so-called because they perform electronically certain operations that were originally in the realm of mathematics and, before that, philosophy. You are probably familiar with the logical exercise that reads:

Socrates is a man.
All men are human.
Therefore, Socrates is human.

Here, A is true and B is true; therefore C is true. This is the philosophical equivalent of what, in electronic logic circuits, is called an AND gate (Fig. 6-1). If input A and input B are both at a positive voltage level (''true''), the result is also positive, or true. If one or the other of the inputs is at ground potential (''false''), the output is also false. Both the A *and* the B inputs have to be true for the output to be so, hence the name of this type of device. In TTL logic circuits ''true'' is represented by a voltage close to five volts, and ''false'' is represented by a voltage close to ground potential. The ''true'' voltage level is frequently called a *logic high*, and the ''false'' one a *logic low*. This system is often referred

Fig. 6-1. A two-input AND gate (left) delivers a logic-high output only when both of its inputs (A *and* B) are also high. The NAND gate at the right goes *low* when both its inputs are high. The ''N'' in ''NAND'' stands for ''negative,'' as does the small circle at its output in the schematic symbol.

to as ***digital logic***, referring to the binary digits the voltages are frequently used to represent.

Other elementary logic circuits are NAND, OR, NOR, and XOR. The letter "N" in the name stands for "negative" and indicates that the output of the gate is the opposite of the input. If the two inputs of a NAND gate are positive, for example, its output will be negative. (The X in XOR, by the way, stands for "exclusive." That's another story, though.) Simple logic circuits can be combined to form complex ones, all the way up to Cray-3 supercomputers.

The outputs of digital logic circuits, although they may be able to source, or output, only a few milliamperes of current, are often used to control higher-power circuits through the use of transistor or other semiconductor-device interfaces.

LOGIC FAMILIES

You can't just go into an electronic parts store, walk up to the counter, and say, "I'd like half-a-dozen two-input NAND gates, please." You have to be more specific than that.

Integrated circuits that perform similar or identical functions are available in a number of different types and styles. Sometimes they bear similar designations and sometimes not. Slight differences in component designations mark differences in the way the ICs function, their suitability for specific applications, and their price.

TTL Logic

The most common logic family is TTL (for Transistor-Transistor Logic). The term "TTL" refers to the design and construction of the logic circuits on the ICs, which consists mainly of transistors connected directly one-to-the-next (hence "transistor-transistor"). Logic families that came before TTL included RTL (Resistor-Transistor Logic) and DTL (Diode-Transistor Logic). While you can still find RTL and DTL parts available, they are there to replace parts from those logic families that are already in use and that fail from old age or other causes. Circuits designed today do not use RTL or DTL ICs.

TTL logic ICs are intended for operation from five volts dc. This supply must be extremely well regulated—within 0.25-volt—and free of spurious noise and other electrical impurities. Although it is frequently assumed that the logic-level input and output signals used among such ICs must be at the five-volt level as well, this is not the case. TTL signal levels can be as high as 0.8 volts for a "logic low" and as low as 2.4 volts for a "logic high." Indeed, these signals rarely reach the extremes of five volts or ground potential that are typically assumed. Remember this when you're troubleshooting logic circuits—what at first appears to be an out-of-range voltage may be, after all, right on the mark.

The designations of all ordinary TTL ICs begin with the number 74. For this reason, TTL logic is sometimes called the 7400 ("seventy-four hundred," not "seven thousand four hundred") series, or 74*xx*. Within the 7400 family are several subfamilies. They are indicated by one or two capital letters immediately following the "74" prefix. The *xx*—two or three numbers that follow the "74" or the one- or two-letter subfamily designation indicate the precise function of the IC. No matter what subfamily it belongs to, the function of a 7400-series IC can always be identified by these last digits. A 7447, 74LS47, and 74H47 all do exactly the same thing, although each type is optimized for certain conditions. Table 6-1 shows the major TTL-logic subfamilies, along with a brief explanation of what each does best.

Table 6-1. TTL Logic Families.

7400	Standard TTL
74L00	Low-power TTL
74H00	High-speed (and high-power) TTL
74S00	Schottky (fast action) TTL
74LS00	Low-power Schottky TTL
74F00	Fast TTL

For the purposes of general experimentation, a member of just about any 74*xx* family will do. It doesn't matter whether the device is low-power, extra-fast, or anything else. All that counts is that it perform the particular logic function you have in mind. This can be a particularly reassuring thought to hold onto if your budget is limited and you have to depend on parts scrounged from surplus circuit boards or from the bottom of someone's junk box.

CMOS Logic

TTL integrated circuits are the mainstay of logic-circuit design, but for some applications they leave many things to be desired. For one, they require a very stable source of five volts dc, something that is difficult to guarantee in a portable or automotive situation. (It is also inconvenient to get an output of five volts from any type of readily available battery pack. The output voltage is always too high or too low.)

Furthermore, TTL ICs are extremely susceptible to electrical noise—spikes and other "glitches" on their power supply and signal lines. The ICs tend to confuse this noise with the signals they are supposed to be processing, and the result is nonsense, frequently referred to as "garbage."

Integrated circuits fabricated using CMOS technology overcome these difficulties. The term "CMOS" stands for "Complementary Metal-Oxide Semiconductor" and CMOS ICs are the integrated-circuit equivalent of FETs. While CMOS ICs are, like FETs, more susceptible to damage from static electricity than are TTL circuits, once they are in place their benefits make them shine.

The supply voltage requirements for CMOS ICs are much less stringent than they are for TTL ones. CMOS supply voltages can range from about three volts dc to as high as 18 volts. The higher the voltage, the faster CMOS ICs can be operated, although the heat they generate also increases with voltage and can become a limiting factor in some instances. CMOS ICs are also relatively immune to electrical noise, which makes them especially well suited to use in such places as automobiles, where spark plugs act as miniature spark-gap radio transmitters.

It used to be that CMOS circuits were considerably slower than their TTL counterparts, but in recent years significant improvements in speed have been achieved.

Most CMOS integrated circuits belong to the "4000" family. That is, their designations appear as "40*xx*." For some reason, CMOS ICs from Motorola are designated 140*xx*. If you ignore that first "1," you'll have the industry-standard part number.

The significant digits (the *xx*) used to identify CMOS ICs are not interchangeable with those used in TTL-series ICs. A 4047 integrated circuit does not perform the same function as a 7447!

Other Logic Families

Not only are there differences between TTL and CMOS IC part numbers when it comes to functions, but the pinouts for ICs with similar functions are also different. Pin five on a CMOS logic IC of a particular sort does not serve the same function as pin five on a TTL IC that does the same thing. Never assume anything about the pinout of an IC—always refer to the data sheet.

To overcome the incompatability problem, particularly for situations where a designer might want to replace TTL ICs with CMOS ones for reasons of noise immunity or lower power consumption, a new family of logic ICs was introduced: the 74C*xx* series. The "C" refers to the fact that these ICs are fabricated using CMOS technology and offer all the benefits of CMOS. The 7400 number means that they are functionally identical to TTL ICs bearing the same designation, and have the same pinout as well. The pinout of a 74C154, for example, is identical to that of an ordinary TTL 74154 and the two parts work in exactly the same way.

As in TTL parts codes, letters that appear in the middle of the designation describe special characteristics of that subfamily. Table 6-2 shows some of the variations on the 74C*xx* theme.

Table 6-2. 7400-CMOS Logic Families.

74C00	Non-standard pinout, function corresponds to TTL part number
74HC00	High-speed, non-standard pinout, function corresponds to TTL part number
74HCT00	High-speed, TTL pinout function corresponds to TTL part number

While the idea of using similar designations to designate ICs performing the same function and having the same pinouts is a good one, it seems to have come a little late in the game. Designers who choose TTL because of its performance characteristics know exactly what they want and can reel off part numbers and functions in their sleep. CMOS advocates can do the same for their favorite ICs. The two families have become so entrenched in their individual niches that many engineers are uncomfortable treading on that middle ground where the part number from one family designates a component from the other one. In your own work, you will probably find it useful to stick to one branch or the other, and leave that crossover ground for the rare instances when you find you have to substitute a CMOS component for a TTL one without changing the surrounding circuitry.

CMOS AND STATIC ELECTRICITY

Static electricity—large accumulations of electrons in one place (as opposed to the flowing electrons of an electric current)—is a big enemy of metal-oxide semiconductor devices such as FETs and CMOS ICs. The delicate junctions in them can easily be zapped—destroyed by a high-pressure rush of electrons rendering the devices useless. Several types of precautions are taken by the manufacturers and vendors of these damage-prone components to ensure that they reach their destinations safely.

Field effect transistors are often shipped with a small twist of fine wire about their leads. This wire, which of course has to be removed before the component is used, sees to it that any electrical charge that may be applied to one or more of the leads take the path of least resistance—namely the wire itself—to the others. By diverting the electron flow around the outside of the transistor, as it were, and seeing to it that each element of the device is equally charged, the wire prevents the sort of destructive damage that static electricity might otherwise cause.

CMOS integrated circuits, being constructed along the same principles as FETs, are also vulnerable to damage from static electricity. Even though many of them incorporate so-called protective diodes intended to block current flow in the wrong direction, large voltage differentials between the pins of a CMOS device can still result in catastrophic damage. To prevent this sort of harm, shippers of CMOS devices use a number of ways to neutralize static electricity.

ICs are frequently shipped in tubes capable of holding a dozen or more devices. Although sometimes made of metal, these tubes are usually made of a conductive plastic, often (although not always) pinkish in color, that serves the same function as the wire around the leads of an FET. Being a conductor, the plastic prevents buildups of static electricity at any one point on the tube. And, if the tube itself is exposed to a jolt of static electricity, the material conducts the electrons equally to all the pins of the ICs contained within the tube, preventing internal damage to them. You might also come across conductive plastic in the form of "bubble wrap" or the "peanuts" used to cushion and protect items such as circuit boards when they are shipped.

Individual CMOS ICs are often shipped with their leads stuck into pieces of conductive black foamed plastic. This material, usually just called conductive foam, has a distinctive feel and appearance and is quite different from the white (or otherwise colored) Styrofoam sometimes used for shipping or storing less sensitive types of ICs. The conductive foam serves two purposes. First, it keeps the pins of the IC from being bent in transit, very important when it comes time to mount the device. And, second, it neutralizes the effects of static electricity just as conductive shipping tubes and the FET wire do.

When you receive a shipment of ICs or other static sensitive components, do not discard the protective shipping material. Use it to store the devices until they are used, and reuse it for others.

Working Around Static Electricity

Static electricity is a serious problem in many parts of the country at various times of the year. It becomes especially troublesome when the humidity is low—in winter in the East and during the summer Santa Ana winds on the West Coast. Some parts of the country, such as New Mexico, suffer from (or enjoy, depending upon your temperament) low humidity all year round.

We've seen how static electricity problems can be avoided in shipping and storing sensitive semiconductors, but what do you do when you're ready to take these devices out of their protective environment and use them? There are many commercial solutions available if you care to spend the money for them—items such as grounded wrist straps to keep you static-electricity free, and grounded soldering irons to keep charges from being transferred from one place to another. Until you go into large-scale big-time production, though, simpler precautions will suffice.

The first, and most elementary, is to keep static charges from building up on yourself in the first place. This means that your work area should have an uncarpeted floor and that you should wear clothing made of materials such as cotton that do not generate static electricity the way synthetic fabrics such as nylon do. Also, don't pet the cat while you're working.

It's a good practice to discharge yourself before sitting down to work with static-sensitive devices. Computer and office-supply stores sell spray cans of a fluid that is supposed to dissipate static charges on such things as carpets. If you have to work in a carpeted area, try spraying some of this where you walk and work. While this spray is somewhat effective, it wears off quickly and has to be reapplied often.

You can make a simple discharging device for yourself from a metal plate a couple of inches square connected by a length of wire to a good ground—say, a cold water pipe. Even a hot water pipe (a radiator, for example) will be better than nothing. Anything, as long as it can "sink" a large number of electrons. A rule of thumb to use is if you can get a shock from touching it after scuffing your feet on the carpet (or petting the cat), then it will do the job. Get in the habit of giving this plate a slap every time you sit down to work. The slap will, of course, cause any excess electrons you may be carrying around with you to be discharged to ground.

As a further precaution, when working with static sensitive devices, you can artificially increase the humidity of your working environment.

In some places a small humidifier is enough. A good steamy bathroom is an excellent place in which to work.

Once components have been mounted on a board and interconnected, the possibility of static-electricity damage is greatly reduced. It is still possible, though, that a jolt can be applied to the board, particularly if it has finger-type contacts at the edge and you find a sensitive input somewhere. When you work on PC boards, observe all the precautions mentioned. Also, cover your working surface with a piece of conductive bubble wrap, or lacking this, with a damp (it doesn't have to be dripping) piece of paper towel, which you can lay on more conventional plastic to protect your working surface. These precautions will equalize any electron concentrations that might build up.

ANALOG DEVICES

The other part of the integrated circuit world is made up of analog devices, also known as *linear* ones. Their inputs and outputs are real-world ones, signals that can be plotted and graphed, as opposed to digital impulses that either are there or are not. Analog devices perform ''conventional'' functions such as amplification and frequency generation—the sort of things you always expected electronics to do before digital circuits entered the picture. Some complex special-purpose linear ICs contain both analog and digital circuitry.

It is in linear ICs that you can really see how integrated circuits have become building blocks in electronic design. For example, let's say that you want to design a regulated five-volt power supply. In the ''good old days'' you would have spent part of your design time in working out the way in which you would do the actual voltage regulation—how you would compensate for variations in input and in load characteristics. Today what you have to spend for all that is about $1.25 for a little three-terminal regulator IC. You feed the power in one end, ground the middle, and out comes five volts, or twelve volts, or whatever it was you wanted in the first place.

The same goes for audio amplifiers. A little ''black box'' IC with a few external components attached to control gain and maybe add some frequency compensation will do it all.

Want to build a TV set? You can purchase a TV fine-tuning circuit, sound-IF amplifier, video-IF system, chroma system and so forth, each on a minuscule chip of silicon encased in a little plastic package. TV manufacturers have reduced this even further for their own use to just two or three ICs.

The list of linear ICs goes on and on. If someone can conceive of a way to put a circuit onto an IC and if there is enough of a demand for that circuit, it will go into production. It is way beyond the scope of this book, of course, to detail the use of these special purpose ICs; we'll stick to the simpler ones here. When you need the fancier building blocks, you know where to find them and you'll know how to use them. In the chapters that follow we'll concentrate on just three types: voltage regulators and operational amplifiers, and a special-function linear IC—the 555 timer—which you'll encounter in Chapter 9.

Voltage Regulators

Until the advent of 7400 series TTL ICs with their stringent power requirements, regulated power supplies were not very common. They were somewhat complicated to design and build, which made them expensive, and were largely found only in scientific laboratories and in other environments where cost was no object but accuracy was. If you needed a power supply that was fairly stable, you overengineered it so that even when you were drawing as much power from it as you might ever need, you still had capacity in reserve. This practice of overdesign ensured that you never pushed your power supply to the point where it had to strain to keep up with your demands—and thus fall off in performance—and also allowed the components of the supply to run coolly and comfortably, extending their life. Overengineering of this sort is still not such a bad idea these days, if you can afford it.

When it became feasible to manufacture entire voltage regulators, complete with voltage-sensing and compensation circuits, on a single piece of silicon, it was possible for everyone to have a regulated power supply for very little more than the cost of an unregulated one.

There are two types of voltage regulator ICs: fixed and adjustable. As you might gather from their names, the fixed regulators deliver a preset voltage, while the output from the adjustable type can be determined by the user.

Fixed output regulators frequently have the part number prefix 78 or 79. The 78*xx* series is intended to provide a positive output voltage, while the 79*xx* delivers a negative one. The last two digits, here shown as *xx*, indicate the voltage. For example, a 7805 regulator delivers five volts, a 7812 delivers twelve volts, and a 7908 has an output of ***minus*** eight volts.

Adjustable regulators do not carry such clear-cut designations. You have to refer to the data books to get what you want (unless you already know).

Operational Amplifiers

Operational amplifiers, more familiarly known as *op amps*, are small-signal devices with a multitude of uses—so many that entire books have been devoted to them. The term ''operational amplifier'' refers to the original purpose of these devices. Way back in the vacuum-tube days, operational amplifiers were used as parts of analog computers. For example, you could use an op amp to multiply by inputting a voltage and amplifying it by a specific amount—that is, by designing the op-amp circuit to have a precise amount of gain. The input voltage represented one of the numbers to be amplified, and the gain represented the other. Their product was the output voltage of the circuit. This amplifier performed the mathematical operation of multiplication, hence it became known as an operational amplifier.

Back in those days, and even later when transistors replaced vacuum tubes, op amps were very delicate and finicky devices, and great pains had to be taken to ensure the accuracy of the calculations. Components had to be extremely carefully matched and selected, and voltages had to be controlled with great precision because they were being used to express numbers that required great accuracy. Early op amps were complicated, bulky, and expensive.

Today an op amp is something that comes in a little eight- or fourteen-pin DIP package and costs a few pennies. This incredible feat—all the more so if you have ever had any dealings with op amps in their earlier incarnations—is, like so much else in electronics today, the result of integrated-circuit technology. It is possible for the characteristics of the components making up an op amp to be matched very closely since they are all fabricated on the same piece of silicon and from the same batch of materials. Similarly, because of the way they are designed and built, and because of their extremely small size, it is possible to maintain the same supply voltage throughout operational amplifier ICs.

Analog computers are used only in specialized applications today; they have been for the most part replaced by digital devices. Op amps, though, are still very much with us. They are extremely useful in small-signal applications, where they find application for such purposes as amplifiers, filters, waveshapers, and as comparators for the conversion of analog signals to digital ones.

7
Integrated Circuit Lore

The small world of integrated circuits is a relatively new one. This chapter covers some of the tricks and techniques that have evolved along with this world, as well as some general IC lore.

CASE STYLES

When it comes to case styles, be it for integrated circuits or transistors, there are more of them than you can shake a soldering iron at. Some designs came about because of engineers' special requirements, while others have evolved from existing types as manufacturing technologies changed or improved. Most of the time, differences in packaging will be immaterial to you—if a part does the job, and you can fit it in, that will be all that matters. There are a few instances, though, where the type of case will be important to you—perhaps to the extent of determining many of the other physical characteristics of your circuit.

Integrated-circuit packages can be plastic, ceramic or metal. Most common, and the least expensive, are those devices in plastic packages. Ceramic ones generally indicate a part with greater latitude with respect to operating temperature and other parameters, and may have been intended for military or other "rugged" use. They are priced accordingly. (The military version of the 7400 family is the 5400-series. The two lines are functionally identical, but the 5400 parts have to pass tougher tests and have a wider operating range.) Round metal cans can contain devices with between eight and sixteen pins. They, too, are generally intended for more rugged environments than their plastic equivalents. Metal cans are often used to afford protection from RF or other electrical interference. If you form the leads of a round metal-can IC carefully with a pair of long-nose pliers, you can use it in an ordinary rectangular IC socket.

Frequently the same device will be available in a variety of package styles. A linear IC such as a 741, for example, might come in an 8-lead metal can, an 8-pin DIP package, and a 14-pin DIP package (Fig. 7-1). "DIP" stands for "Dual In-line Pin (or Package)" and refers to the two rows of pins forming the "legs" of the IC package. The 14-pin-package version of the 741 has several pins marked NC, for "no connection." They are not connected to anything in that package, and are there only for "cosmetic" purposes. Some other devices bring out connections in the 14-pin version that don't appear in the 8-pin one. Several of these pins are intended to be connected directly to other pins of the same IC, something that is done internally in the 8-pin versions.

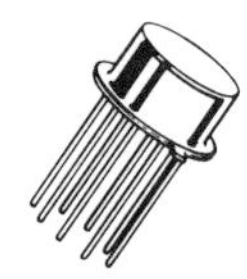

Fig. 7-1. Integrated circuits come in a variety of different package sizes, depending primarily—with the exception of the metal can, which is a "ruggedized" packaging—on the number of pins that have to be brought out. IC DIP packages with 16 or fewer pins are usually 0.3 inch wide; those with more than 16 are usually double that, or 0.6 inch.

If you have the choice between an 8-lead can and an 8-pin DIP package, you will probably find the DIP version easier to work with. The rectangular DIP sockets are much easier to come by than round sockets. DIP packages of up to 16 pins are all the same width, 0.3 inch. Packages with more than 16 pins—24- and 40-pin ones, for instance—are twice that, 0.6 inch wide.

Power transistors and voltage regulators generally come in one of two case styles: TO-220 or TO-3 ("TO," by the way, stands for "Transistor Outline"). These cases are illustrated in Fig. 7-2. Low-power voltage regulators also come in ordinary plastic transistor-type TO-92 cases (Fig. 7-3). The three-lead TO-220 case has a metal tab sticking up from it. This tab is usually connected internally to one of the elements of the device, meaning that it duplicates the functions of one of the leads. The tab can be used for heat-sinking, or sometimes for attaching the device to a chassis or circuit board.

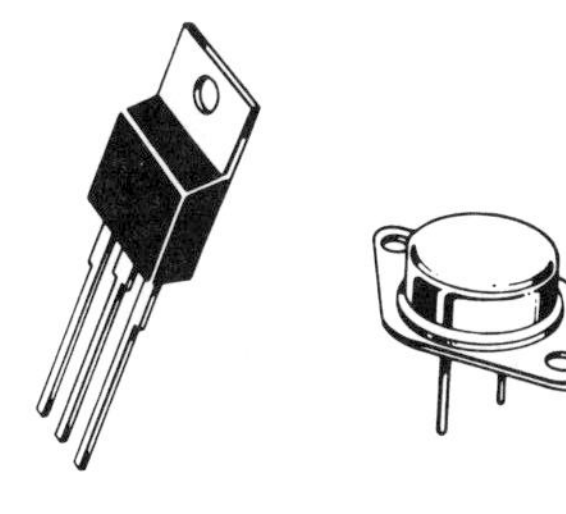

Fig. 7-2. These two case styles are used for ICs, both regulators and other types, handling high currents. The TO-220 package at left is good for more than an amp when properly heat-sinked; the TO-3 package at right has a considerably higher power-handling ability.

Fig. 7-3. Low-power (a few hundred milliamperes) voltage regulators may come in "transistor-type" TO-92 plastic packages.

Heavier duty versions of power transistors and regulators are packaged in TO-3 metal cases. These lozenge-shaped cases should always be chassis-mounted. As with TO-220 devices, one internal element of the device is usually attached to the case, requiring the use of special mounting hardware to prevent the case from shorting out to ground (unless this is desired—which it usually is not). The chapter on power supplies talks more about this matter. Devices in TO-3 cases can handle several amperes if properly heat sinked.

The cases used to package linear integrated circuits also come in several varieties. The case style is usually indicated in a suffix added to the end of the part number. Thus, you can find an LM741H, LM741N, and an LM741D. As illustrated in Fig. 7-4, the -H version is in a metal can, the -N version in a plastic DIP package, and the -D one is intended for surface mounting.

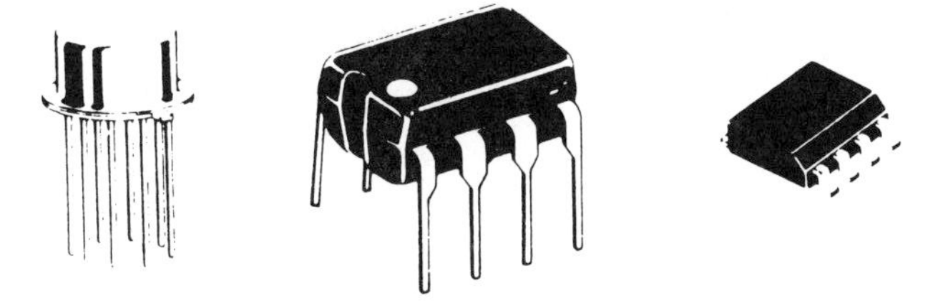

Fig. 7-4. The suffix attached to a part number can indicate the type of package used. A metal can carries the designation "H" (for "hermetic"), an eight-pin plastic package "N," and a surface-mount one "D."

Confusing IC Designations

It's usually not difficult to identify ICs. Their part numbers are printed right on them, usually on the top, but sometimes they're on the bottom, and occasionally on both the top and bottom of the device. There is a situation, though, that occasionally makes it confusing to figure out what type of IC you have.

Manufacturers often stamp IC packages not only with the part number, but also with the date of manufacture (and sometimes other information as well). This date is usually presented in the form *YYWW*, where *YY* stands for the last two digits of the year of manufacture, and *WW* the week of the year. The marking "8701" would represent the first week of 1987, "8752" would be the last week of that same year.

A simple logic gate you found on a surplus circuit board might be marked "7404" . . . twice! One marking would be the part number, the other to indicate that it was made around the end of January in 1974. Such cases are rare now. We're pretty safe now until the year 2040, when the CMOS manufacturers will begin to frustrate us.

The part designation of this device is "7805," indicating that it is a five-volt tab-type voltage regulator. The other number, "7939," tells you that it was manufactured during the thirty-ninth week of 1979. Radio Shack

Surface Mount Devices

The manufacturers, designers, and users of semiconductor devices always want to make things smaller, and to pack more into less space. This applies both to putting things into packages, and to putting packages on circuit boards. The current "rage" is for *surface mount* devices.

The leads or pins of conventional IC packages are intended to be inserted into holes in a circuit board (where they are soldered) or into sockets that are themselves soldered to a circuit board. (For prototyping, these pins and leads can, of course, be inserted into the spring-clip-loaded holes in a solderless breadboard.)

Commercially, ICs are usually inserted into PC boards and soldered there by automatic equipment. And, while the mounting holes drilled in the boards may certainly be a convenience for human assemblers, they are not a necessity for machines. It occurred to some clever engineer that a great deal of the material and space that was being used to mount ICs in the usual manner could be done away with. By eliminating the pins in a conventional IC package, or more properly, by leaving just vestigial stubs of these pins and soldering them directly to the top of a circuit board rather than drilling holes, inserting the leads into them, and then soldering them on the other side of the board, a tremendous saving of metal, real estate, and money could be realized. This same economy could be applied to resistors and capacitors, as well. The result of this thinking was the surface-mount package shown in Fig. 7-5.

Fig. 7-5. Surface-mount packages are used without sockets—they are soldered directly to foil traces on the circuit board. This saves material, space, and money. Copyright Motorola, Inc. Used by permission.

While the pinouts of surface mount ICs are the same as those of their more conventional counterparts, using them in a circuit requires a complete redesign of the circuit board. First, the pin spacing is different. Full-size IC packages use a pin spacing of 0.1 inch. Most printed circuit devices also use this tenth-of-an-inch spacing, as do the solderless breadboards that accept them. Surface mount ICs, however, have their pins spaced only 0.05-inch apart. This twentieth-of-an-inch spacing makes for greater compactness, but it also means that surface mount components aren't compatible with PC-board layouts for standard ones.

Furthermore, standard devices are usually mounted on the side of the circuit board opposite that which contains the signal- and power-carrying traces. (This is why that side of the board is often called the "component side.") Since surface mount devices are intended to be soldered to traces on the same side of the circuit board as that on which they are mounted, the routing of the traces has to be reversed.

It is also unnecessary to drill mounting holes for the parts since they are soldered directly to pads at the ends of the traces to which they are connected electrically.

Surface mount components are really intended to be handled by machines; it takes a steady hand and an eagle eye (or a good magnifier) for people such as us to use them. Still, it is difficult to resist the temptation to squeeze as much as you can into as little space as possible, and you will probably at some point find yourself wanting to use these miniature miniatures. You would be well advised, though, to do your developmental work using standard size components, switching to the small surface mount variety only after all design changes have been made and you are ready to build the final functioning version of your design.

PIN NUMBERING

Back in the days of vacuum tubes there was only one way you could insert a tube into its socket. There was either a "key" on the tube that fit into a slot into the socket—making it necessary to align the two and thus ensuring that all the pins went into the correct holes—or the pins themselves were spaced unevenly to match a similar uneven spacing of holes in the socket. This ensured that the tube could be inserted only one way. You could almost replace tubes in the dark, and a lot of people probably did when fumbling around the back of some crowded chassis trying to replace one tube without having to take the whole thing apart.

Integrated circuits aren't as simple in this respect. With the same number of evenly spaced pins on each side of the package, it's no problem at all to insert an IC backwards into a socket or the holes drilled in a PC board. However, finding the correct orientation for an IC package is less a matter of blind luck than it might seem. The first thing to remember is that the pins on ICs are always numbered counterclockwise, as viewed from the top of the device. Pin 1 is always at the upper left and the last-numbered pin (usually 8, 14 or 16) is at the upper right (Fig. 7-6). It is easy to become confused about this when working on a circuit board from the solder (under-) side. Remember that from this viewpoint, pin numbering goes the other way—clockwise.

The pin-1 end of an IC is marked in several different ways. The mark may be a U-shaped notch molded into the IC case, or it may be a small dot-shaped depression next to pin 1. The pin numbers on ICs that

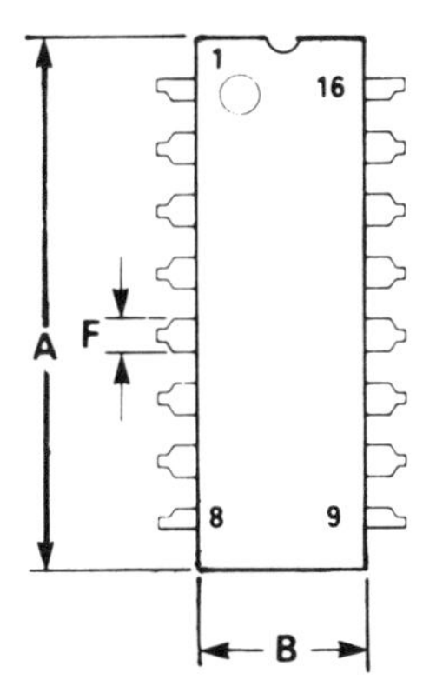

Fig. 7-6. Pin numbering on IC DIP packages starts at the upper left and proceeds counterclockwise—as viewed from the top—around it. Copyright Motorola, Inc. Used by permission.

are packaged in round metal cans also run counterclockwise looking at them from the top. Pin 1 is indicated either by a gap in the regularly spaced leads or by a tab on the metal case. Pin 1 is to the left of the gap or tab.

IC sockets also sometimes have a pin-1 marking, although it is not necessary in their case, merely a convenience to help you in inserting the devices properly. A socket's pin-1 end can be indicated by a corner that looks as though a small triangular piece has been snipped off. Don't be concerned if no pin-1 orientation is apparent on an IC socket; they're symmetrical anyway and there is no "wrong way" to insert a device into them. If the sockets are correctly oriented, the ICs inserted in them will also be.

Snug as a Bug

You can fit two 8-pin DIP packages in a 16-pin socket. This will save a little real estate, the price of a second socket, and maybe a bit of design and layout work.

INSERTING AND REMOVING ICS

One of the strong points of solderless breadboards is that components can be inserted in and removed from them repeatedly, making it possible to reuse expensive devices and to reorganize circuits with little effort. To do this you have to know, of course, how to insert and remove IC packages without damaging them.

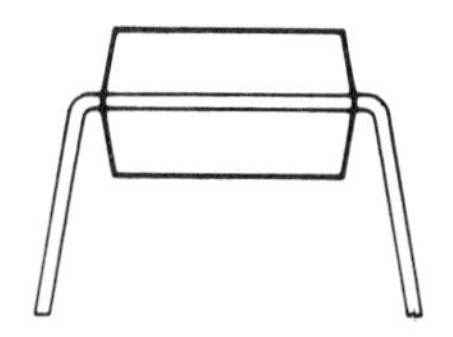

Fig. 7-7. The pins on brand-new DIP packages are usually spread somewhat to accommodate the requirements of automatic insertion machines. The text describes an easy technique for preparing them manually for insertion into breadboards or sockets.

Brand new ICs come with their "legs" spread wide, as is illustrated in Fig. 7-7. This is to facilitate their handling by the automated insertion equipment used in production line manufacturing. To make them fit into sockets or into the holes of a solderless breadboard, the pins must be straightened so they are perpendicular to the IC package and pointing straight down. There are all sorts of tools—pin straighteners and insertion devices—that supposedly make this job easier, if not effortless, but they are not really necessary.

The spread-out pins of a brand-new IC can be readied for insertion by placing the IC on a hard surface such as a tabletop with one row of pins resting on the tabletop and facing away from you. By gently gripping the package at the ends and "rolling" it away from you, you can bend all the pins on the downward side inward simultaneously. When you're done, the pins on one side of the IC package will be at right angles to it. Turning the IC around, you can then do the other row of pins in the same fashion.

It's best to proceed slowly until you get the feel of this procedure; it's not as easy to unbend the pins if you get carried away and go too far. Roll the package a little, check the results, and then roll a little more until things look right. Pins that are still a little bit spread apart will generally present no difficulty, and it's certainly better to have them that way than it is to have them "pigeon toed."

The worst thing that can happen when you're inserting an IC into a socket or other mounting place is to have one of the pins bend. While manufacturers' specifications say that a pin must be able to survive three flexures (bendings) without breaking off, even one mishap of this sort will weaken the pin and make it more difficult to insert into a tight hole without bending again.

The best way to insert an IC into a socket is to do so little by little, from one end to the other, and from one side to the other. In other words, don't try to insert all the pins at once. Start with one corner pin and work your way up the pins on that side of the package. When that entire row is partially inserted (it doesn't have to be fully seated yet) you can insert the other row, usually all the pins almost at once. It may be necessary to apply a little sideways pressure to the package to align the second row of pins with the holes into which they are to go, but this will not harm anything. Just try not to do anything too suddenly, for that may result in an entire row of pins abruptly giving way. When all the pins have been started in their holes, seat the device firmly by rocking it into its socket with gentle pressure at first one end and then the other.

If a pin does bend, remove the IC from the socket, gently straighten the bent pin with a pair of long nose pliers so it is exactly in line with the other pins (this may take a little doing; again, work a little bit at a time), and try once more to insert the device.

All may not be lost even if a pin does break off a valuable IC. Check the pinout of the device—maybe the pin that's no longer there is unused in your circuit, or maybe there is an extra device in the package that's been unused until now and that can be substituted for the one that's now inaccessible.

There are even a couple of ways to reconstruct the missing pin, although they would make most professionals cringe, and take a bit of ticklish work. The first way is to make a ''prosthesis'' for the amputated leg using a piece of resistor lead tack-soldered to the shoulder of the missing pin. After it is soldered in place, the excess portion of the lead can be cut off, and the replacement pin bent into alignment with the others.

It may also be possible to use a piece of fine wire-wrap wire to make the missing-pin connection. Strip one end and wrap a couple of turns around the stump of the missing pin (if there's enough material left to do so), or tack solder the wire to it. Then route the wire to a solder pad that is located on the same trace the missing pin would have been connected to, or to any point you know would have been directly on the circuit path. Tack solder it there. This is an awkward and clumsy way to effect a repair, and the extra length of the wire can prove deleterious in high-frequency circuits. It is certainly not recommended, but it can sometimes be the only way out of an otherwise hopeless situation.

IC Removal

Just as there are special tools for IC insertion, there are also those for removing the packages from sockets. However, you can frequently do without them.

Never try to remove an IC package from its socket with your fingers. You'll find that as you pull and tug nothing happens until . . . all of a sudden the whole thing comes loose at one end, bending all or most of the pins beyond the point at which you even want to try to think of unbending them.

You can, however, gently pry an IC out of its socket from underneath. With the tip of a fine-bladed screwdriver or with the short end of an L-shaped piece of metal carefully begin to work one end of the package loose from its socket. Raise it just enough to slip the screwdriver (or the long end of the L-shaped tool) between the package and the socket and, by rocking and prying, free it. You will sometimes have to work on both ends alternately until there is enough room to slip your prybar all the way underneath the package.

Specialized removal tools that clamp around the body of an IC package and give you, through the tool, a firm grip on it, are handy for removing ICs from tight places where there's no space to maneuver a "crowbar." Use a rocking motion to free the IC gently from the grip of the socket, otherwise you can find yourself with a collection of bent pins. Space permitting, removal tools intended for 16-pin packages can also be used for 14- and 8-pin ones.

STORING ICS

After you've built, rebuilt, and scrapped a number of circuits, you'll have accumulated quite an inventory of components—resistors, capacitors, transistors, integrated circuits, and other items too numerous to mention. While you will undoubtedly come up with your own method for organizing and keeping these parts, you might also appreciate a helpful hint or two.

Just as you can store CMOS ICs in black conductive foam you can use a similar technique to keep your TTL and other DIP-package ICs in order. The foam needn't be of the conductive type—TTL devices are not nearly as sensitive to static electricity as CMOS ones. That's not to say that they're completely immune, but they don't have to be treated with as much care in this respect.

Plain white Styrofoam is good enough for TTL. You can get great slabs of it from any florist—they use it to prepare flower arrangements, to hold the stems in place. You can plant your ICs in it. Not only will it keep them organized, but with their leads safely embedded in white plastic they will be guarded from the sort of bending and other damage that they would otherwise encounter at the bottom of your junk box.

PART TWO

8
Power Supplies

Power supplies are almost ridiculously simple to build—there is nothing very critical about them and they can be slapped together just about any which way. If you can't obtain a component of the precise type or value specified, come as close to it as you can. You probably won't notice the difference.

A FIVE-VOLT POWER SUPPLY

TTL integrated circuits require tightly regulated power supplies capable of supplying five volts of very pure dc with virtually all the ripple or hum filtered out, to an accuracy of five percent. That means that the output voltage of the supply cannot fall below 4.75 volts, nor must it rise higher than 5.25 volts, no matter what type of load is place on it—that is, no matter how much or how little current it is called upon to deliver at any instant. Regulated power supplies that could compensate for variations in load used to be a nuisance to design. Today, however, the availability of three-terminal voltage regulator ICs makes them a snap. The five-volt supply we'll describe here will supply about an ampere of current—more than enough for the few IC or transistor packages you'll use it with for the projects in this book. This supply will also be invaluable should you start experimenting the TTL logic circuits.

There is nothing in the design of this power supply that can't be tinkered with; the circuit is presented, as are all the circuits here, more as an example of how to design a power supply than as the ideal one. You can even use any color wire you want—it isn't going to affect the way the power supply works.

TRANSFORMERS

The purpose of a transformer in a power supply is to reduce the voltage input to the supply—which is usually somewhere around 120 volts ac—to a level commensurate with that you want from the output of the supply. When you go shopping for a transformer, you will usually find it specified as being rated at so many volts at such-and-such a current: say, 12 volts at one ampere. Those are the values of its secondary winding, the output side of the transformer. The input you know will be 120 volts. (And, although it is not significant here, the current through the primary will be in inverse proportion to the ratio of the primary to the secondary windings. That is, in a 12-volt transformer, that ratio will be 10:1, or 120 volts : 12 volts. The current through the primary of the transformer, as illustrated in Fig. 8-1 will be one tenth that flowing through the secondary; if the secondary is putting out an ampere of current, the current in the primary winding will be one tenth that, or 100 milliamperes.)

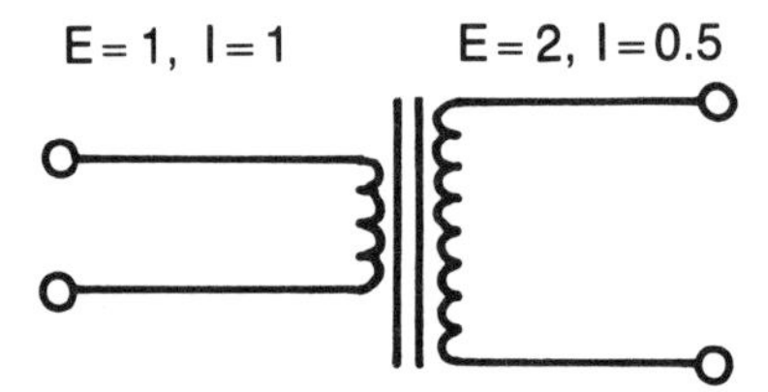

Fig. 8-1. If the number of turns making up the secondary winding of a transformer is double the number making up the primary winding, the voltage in it is doubled, while the current flowing in it is halved. Other windings ratios affect the voltage and current in the secondary similarly.

Many power transformers (there are other kinds of transformers as well—for coupling amplifier stages, for tuning IF stages in radio equipment, and for other purposes) have multiple secondary windings to output several voltages simultaneously (Fig. 8-2). For a simple five-volt power supply you need only a single output voltage. The secondaries (output windings) of power transformers come in two types: center-tapped and non-center-tapped (Fig. 8-3). The first type of power supply we'll look at uses the more straightforward design, the one without the center tap. Afterward we'll consider a power supply that uses a center-tapped transformer.

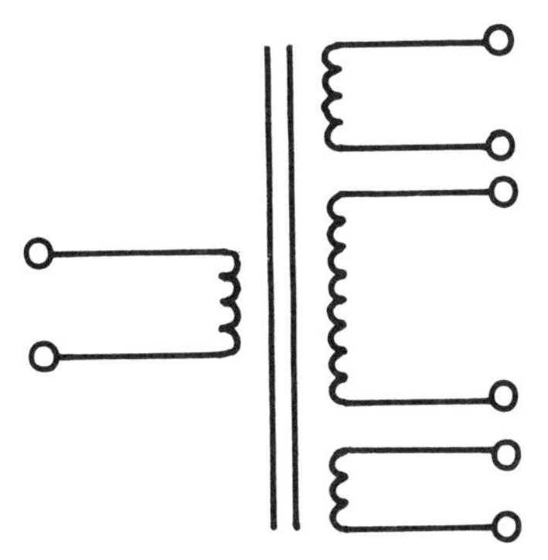

Fig. 8-2. Transformers with several sets of secondary windings, while not required in many simple solid-state devices, are often found in apparati such as computers and TV receivers, where several different voltages are called for.

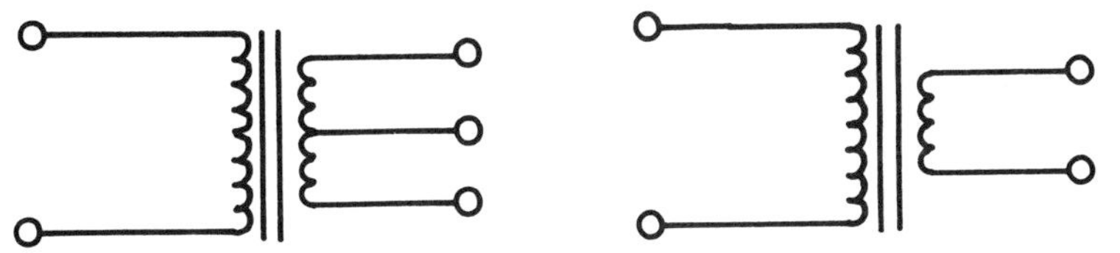

Fig. 8-3. A transformer with a center-tapped secondary winding (left) is particularly useful in constructing bipolar power supplies, those outputting both a positive and negative voltage. Of course, the center tap can be ignored if you like, and the transformer treated just like the non-center-tapped one shown at the right.

If all you can find is a center-tapped transformer, just use the leads from the ends of the secondary windings and ignore the one from the center tap. If the transformer is one of those where the two halves of the secondary have their leads brought out separately, just connect the two from the center of the winding together to make one long winding.

The value of the secondary of the transformer should be about $^2/_3$ the desired output voltage of the supply, although a little more is permissible if a larger transformer is more readily available of if you can find one at a good price. The main thing is to ensure that the input voltage to the voltage regulator is at least two volts more than its output. For this five-volt supply, the regulator input voltage should be at least seven volts. If the input-output difference is less than about two volts the regulator might not work properly and the output voltage will not be stable—and instability is one thing you don't want in a regulated supply.

How can the output voltage of the transformer be less than the output voltage of the supply? There are several reasons. First, the output of

An Easy Center-Tapped Transformer

If you need a center-tapped transformer with a particular voltage or current rating but are unable to locate one, you may be able to take advantage of a simple way around the difficulty. You can use two identical transformers, each of whose outputs is half the voltage of the center-tapped unit originally called for. Connect the primaries of the transformers in parallel and the secondaries in series, as illustrated in Fig. 8-18. This will produce the equivalent of a center-tapped transformer whose primary has twice the current-handling capability of one transformer alone, and whose secondary is "center-tapped" at the point where the two individual secondaries are connected.

If you find that the output of your homebrew center-tapped device is almost nonexistent, you have connected the secondaries of one of the transformers backwards. Its output is 180 degrees

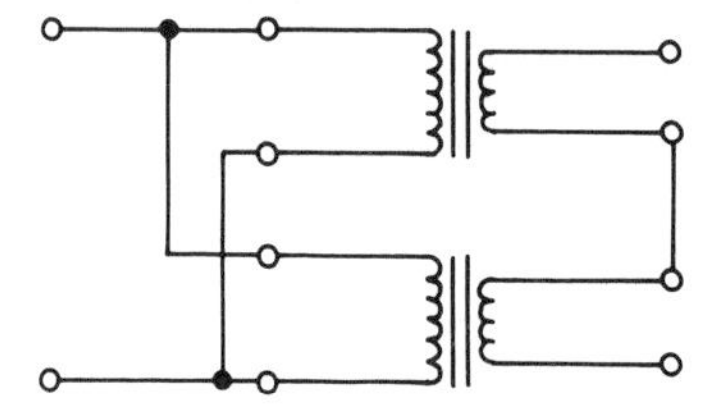

You can make your own center-tapped transformer by connecting the primary windings of two identical non-center-tapped ones in parallel, and the secondary windings in series. The center tap is taken from the point where the two secondaries are joined.

out of phase with the ouput of the other transformer, and the two are canceling each other. Reverse the leads of one of the two transformers' secondaries and things will straighten themselves out.

Incidentally, you can use this principle of cancellation to construct transformers with unusual output voltages. The output voltage of two transformers whose secondaries are connected out of phase will be the difference of the two. For example, a transformer with a 12.6-volt secondary connected out of phase with a 120-volt isolation transformer (whose output is the same as its input—it is used to provide a degree of electrical isolation between its primary and its secondary) would yield an output of 120 − 12.6, or 107.4 volts. The transformer that produces the difference voltage is called a bucking transformer.

the transformer is measured in volts RMS. "RMS" stands for "root-mean-square" and refers to the way the voltage of the ac waveform is measured. The ac waveform is that of a sine wave, and it varies over time—in the U.S. sixty complete cycles per second. We could measure the peak voltage of the waveform and use that, but for most purposes that value would have little meaning. Instead, the RMS value, which closely approximates the equivalent dc voltage of the ac waveform, is used. The RMS value is determined by measuring the voltage of the waveform at a number of different points, squaring each value, taking the average (mean) of these squares, and then taking the square root of that average. In the end, it turns out that the RMS value of a sine wave is 0.707 times its peak value and, conversely, the peak value is 1.414 times the RMS value. The 6.3-volt secondary of the power supply transformer in Fig. 8-4 has a peak output of almost nine volts (1.414 × 6.3).

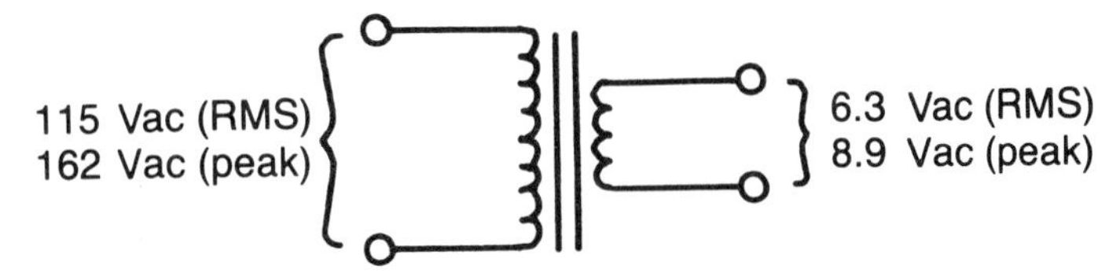

Fig. 8-4. The RMS value of an ac waveform is a way of expressing its equivalent dc voltage. The peak voltage of the waveform is 1.414 times the RMS voltage.

RECTIFIERS

The first step in turning ac into dc is rectification. The term comes from Latin, and means "to straighten out" which, when you come to think of it, is just what you want to happen—you want to straighten the sine wave of the alternating current into a line of constant amplitude representing direct current. The component that's used for this process is a diode, which permits current to pass in one direction while blocking its flow in the other; right away you can see that this property will affect alternating current, in which the direction of current flow is constantly reversing.

TYPE	*PIV*
1N4001	50
1N4002	100
1N4003	200
1N4004	400
1N4005	600

Figure 8-5*A* shows the action of a single diode on an ac waveform. The result is called half-wave rectification—the diode acts on only one half the waveform. Its rather rough output is known as pulsating dc, and is not really suitable for a power supply that has to deliver very clean dc.

The full-wave rectifier in Fig. 8-5*B* acts on both the positive- and negative-going halves of the waveform. Its output is a lot closer to the dc "straight line" ultimately desired. A full-wave rectifier of this sort must

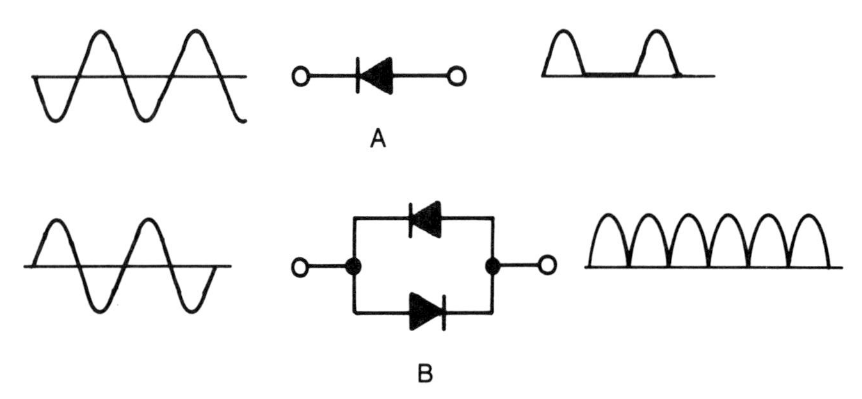

Fig. 8-5. The output of a single diode half-wave rectifier (*A*) is a rather rough waveform called "pulsating dc." The two-diode full-wave rectifier in (*B*) produces a waveform, also called "pulsating dc," that is "fuller" and much easier to filter into true direct current.

be used with a center-tapped transformer, however, and since the two halves of the transformer's secondary winding are treated independently, the total voltage developed across the secondary must be about twice that of the non-center-tapped transformer that would be used in a half-wave-rectified supply.

The best of the two worlds is the four-diode device shown in Fig. 8-6, called a full-wave bridge rectifier. The term "bridge" refers to the configuration of the four devices. There are, for example, such things as resistance bridges, with four resistors connected in a similar four-leg arrangement. A bridge rectifier operates on both halves of the ac waveform and does not require a center-tapped transformer. Its output is much easier to filter than that of a half-wave device. The output voltage of the transformer can be the same as that used in a half-wave rectifier circuit.

Figure 8-7*A* shows how a full-wave bridge is used with a non-center-tapped transformer to produce a single output voltage. In Fig. 8-7*B* a full-wave bridge is used with a center-tapped transformer and connected

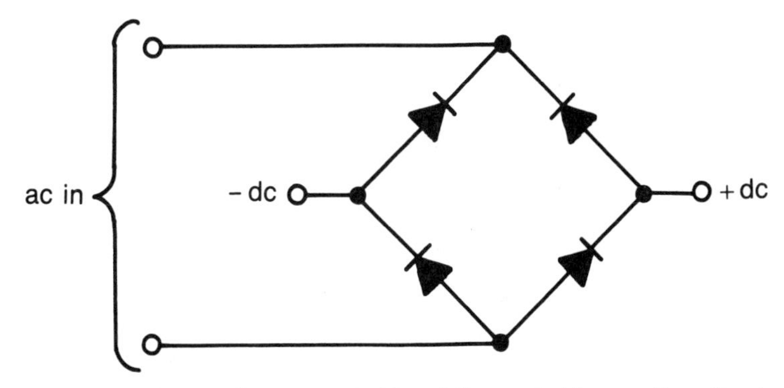

Fig. 8-6. Four diodes can be connected in a full-wave bridge configuration to provide very good full-wave rectification.

somewhat differently, to form the basis of a bipolar power supply outputting both positive and negative voltages. The remainder of this type of circuit will be shown below.

Bridge rectifiers can be made from four discrete devices, or are available in a single package with four leads brought out (Fig. 8-8). The two ac inputs are usually marked with a ~ symbol (it doesn't matter which of the two leads from the transformer's secondary goes to

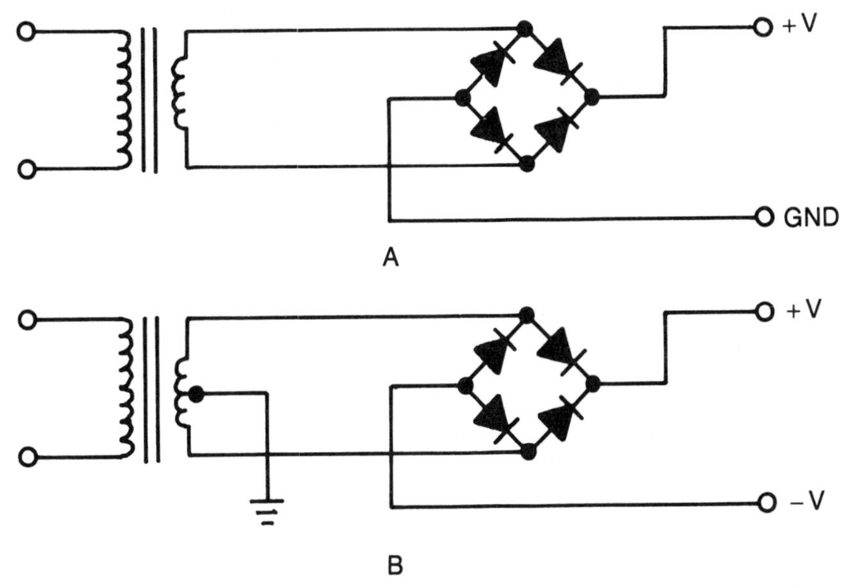

Fig. 8-7. A full-wave bridge rectifier can be used to provide either a single-voltage output referenced to ground (*A*), or a bipolar (plus-and-minus) output if it is used with a center-tapped transformer (*B*).

Fig. 8-8. Although you can make your own from four discrete rectifier diodes if you are so inclined, full-wave bridge rectifiers are conveniently available as single-package devices. The "AC" on the case refers to the two ac inputs, while the "+" and "−" show the polarities of the outputs.

which), and the outputs with signs indicating their polarity. It's usually almost as cheap, and certainly simpler, to use a single-package bridge than to construct one yourself.

Filtering and Regulation

When an ac waveform is rectified into pulsating dc, its value falls somewhere between its RMS and peak values. In the power supply shown in Fig. 8-9, the action of the large capacitor, C1, after it has charged up initially, tends to keep that voltage elevated. It remains at a more-or-less constant level, buoyed up by the large reservoir of electrons making up the charge on the capacitor. The input to IC1, a 7805 voltage regulator, therefore remains fairly constant and is more than two volts greater than the five-volt output we expect from it. The input

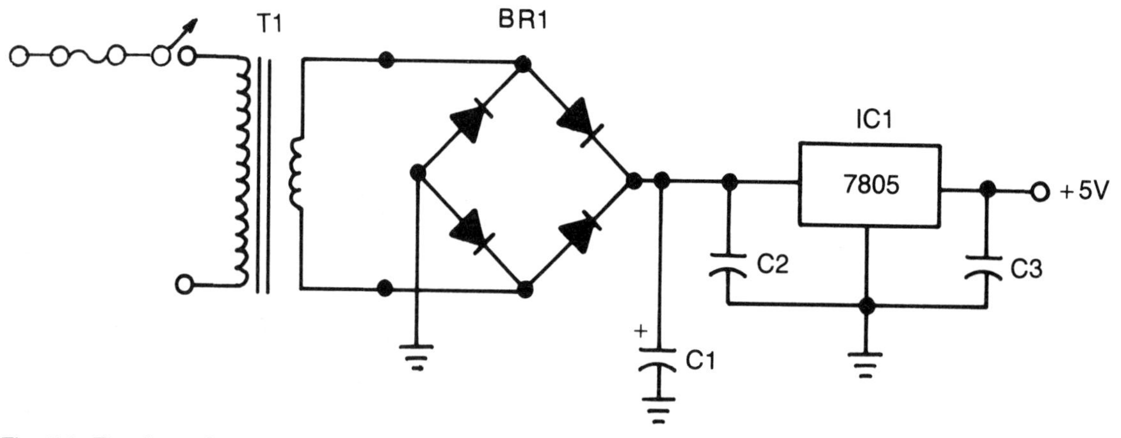

Fig. 8-9. The five-volt power supply referred to in the text.

capacitor should have a value of several thousand microfarads, although you can get away with input capacitors having values of only a few hundred. It depends on how much regulation you want (a larger capacitor smooths out input fluctuations better) and how clean you want the final dc to be (a larger capacitor will also provide a cleaner waveform with less ac ripple, which may be heard as hum in audio circuits). If the regulator will be located some distance away from C1, add capacitor C2, which should have a value of about 0.33 microfarads at its input. That will remove any ''garbage'' that may have crept in. The working voltages of all the capacitors should be at least 50% greater than the voltage they will be handling to provide an adequate margin of safety. A working voltage even higher than that won't hurt, but a lower one can eventually be responsible for the power supply's failing.

While the input voltage to the regulator can exceed its output by more than two volts, it is not a good idea to exceed this difference by too much. The two-volt differential is a necessity, but anything more than that will just be dissipated as heat by the regulator. Aside from being wasteful, this heat can also be damaging to the regulator if there is too much of it, and it can lead to the deterioration of other components in the circuit if they are allowed to get hot.

It is a good idea, if you are going to draw anything near the regulator's rated current from it, to use a heat sink to remove excess heat from it. If the regulator is chassis- or case-mounted, that metal can remove some of the heat. Better, though, is the use of a proper heat sink that is specifically intended to absorb heat from the regulator and transfer it efficiently to the air (Fig. 8-10). You can check the operation of the regulator (and other semiconductor devices that dissipate power in the form of heat) by putting your finger on it. If the device gets too warm to allow you to hold your finger there for more than a second or so, it is running too hot and should be heat sinked to cool it down, and allow it to operate more conservatively.

The small capacitor, C3, at the output of the regulator is there to bypass to ground any small ac component (ripple) that is still present in the dc waveform. This capacitor also helps to stabilize the action of the regulator. A value of 0.01 μF will work well.

You can use the same principles—and some of the same parts—to construct a 12-volt supply. You will probably have to use a transformer with a 12-volt secondary and accept the extra heat that will be generated by the regulator as a result of exceeding the two-volt differential. And, instead of using a five volt regulator such as a 7805 you will, of course,

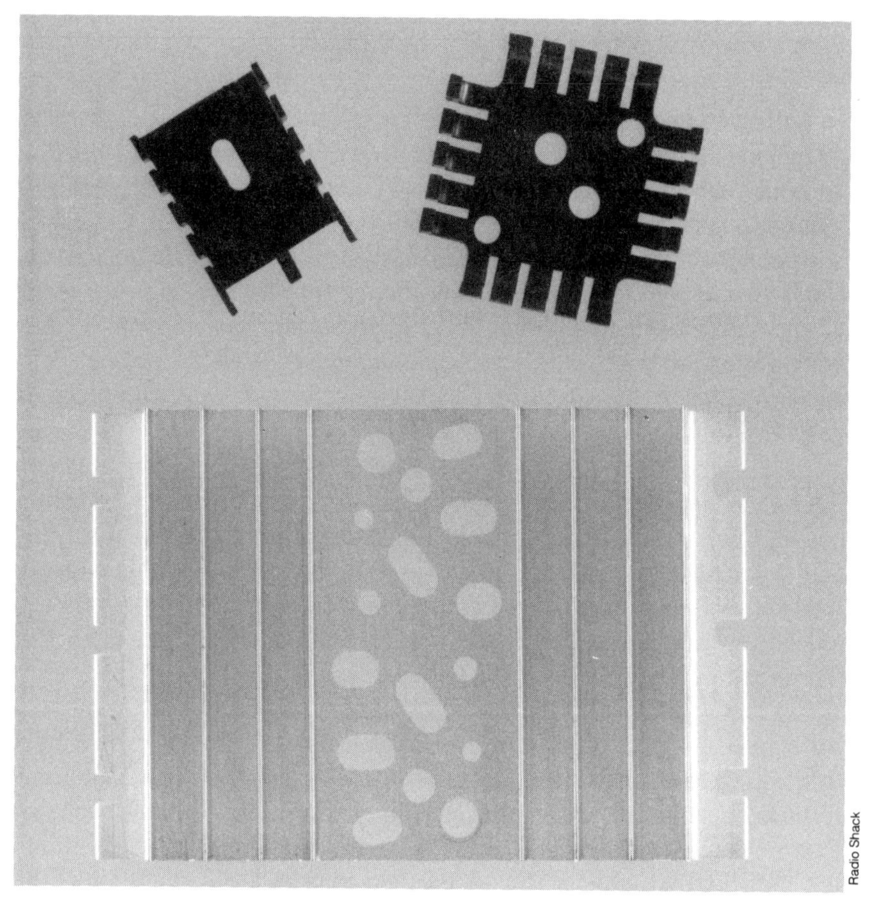

Fig. 8-10. Heat sinks for use with voltage regulators and power transistors. The one at top left is for use with TO-220 packages and the one at the top right for TO-3 packages. The one at the bottom of the photograph is a "universal" type that can accommodate several different lead configurations (and you can always drill more holes yourself if you need them).

use a 12-volt 7812. If the working voltages of the capacitors were already high enough, that's all you'd have to change.

Fuses and Switches

Since the best laid plans of mice, men, and electronics experimenters often go astray—meaning that you never know when something is going to go wrong—it is a good idea to use a fuse or circuit breaker to protect your power supply or the device into which you have built it. Short circuits are not the only thing fuses can protect you from. They can also save you from damage caused by other situations where too much current is drawn. And, while IC voltage regulators are supposed to shut down in the event of short circuits across them, this still does nothing to protect the power supply elements (such as the transformer) that are located ahead of them electrically and that suffer momentary overloads.

Fuses come in three types: fast-blow, normal, and slow-blow. These names indicate how quickly the device will react to an overload. For light-duty devices such as you will probably be designing, fuses with a "normal" rating will do nicely. If, however, your circuit will draw a heavy current initially before settling down to work (this might be the result of charging up a big, big capacitor) a slow-blow fuse, which has a built-in tolerance for temporary current surges, will be the better choice.

Fuses come in several voltage ranges. The ones you'll find will probably be rated at 125 or 250 volts. It doesn't matter which you use, as long as the rating of the fuse matches or exceeds the voltage that will be applied to it—about 120 volts in the case of house current.

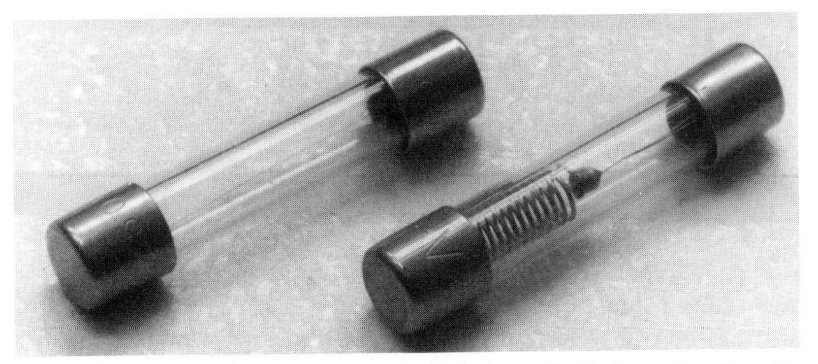

The slow-blow (or "slo-blo") fuse at the right is easily distinguished from its fast-blow counterpart by the details of its internal construction.

Designations such as "AGC," "AGX," and "MDL" indicate the fuse type and case style. What's important to you is the current rating of the fuse, which may be specified separately. You have to calculate approximately how much current your circuit will draw, and choose a fuse value a bit above that. In some circuits you can arrive at this figure by adding up the rated power consumptions of all the IC packages used. At other times, as in the case of a power supply intended to supply an ampere of current comfortably, it is this intended maximum you want to consider.

Since a short circuit, or other catastrophic situation, will usually create a demand for much more current than normal, you have plenty of leeway in determining the value of the fuse you use. Just make sure that it can comfortably handle all the current that the circuit of which it is part will normally supply.

As shown in the power supply back in Fig. 8-9, the fuse should be located in one side of the ac line just where it enters the circuit, before the power switch. This will ensure that it protects *everything* in the circuit . . . even switches can go bad now and then.

As with fuses, switches are rated in terms of the voltage and current they can handle. And, again, as long as the switch can meet the maximum demands of your circuit, you can use it. Electrically, the switch should be located between the fuse and the transformer. This will allow you to shut things down completely—including the primary of the transformer, which is sometimes left "live" in less-thoughtfully designed circuits than yours. Even so, when you work on your circuit, unplug it! Otherwise there will still be line voltage present in the area around the fuse.

A BIPOLAR POWER SUPPLY

A twelve-volt bipolar power supply is shown in the schematic in Fig. 8-11. This type of supply is useful in powering devices such as those using op amps (Chapter 11), which require both positive and negative supplies. The origin of the term "bipolar" comes from the prefix "bi" (two) and the "polar" in "polarity."

This supply requires a center-tapped transformer, or two identical transformers connected "back-to-back" to form the equivalent of one center-tapped one. The output voltage across the entire secondary of the transformer must be twice the output voltage you desire—instead of 12.6 volts, for example, it would have 25.2 volts. This is because you will be considering the center-tapped secondary as the equivalent of two transformers, each of course, with an output of 12.6 volts. Similarly, the current rating of the secondary has to be twice what is required by a single side of the power supply—again, because the single transformer is going to do the work of two.

The 25.2-volt center-tapped transformer and diode bridge are connected so the bridge actually works as two, two-diode, full-wave rectifiers. Compare this with the bridge hookup in the single-polarity supply shown in Fig. 8-9. The positive and negative halves of the supply are constructed similarly, the difference being the the regulators used.

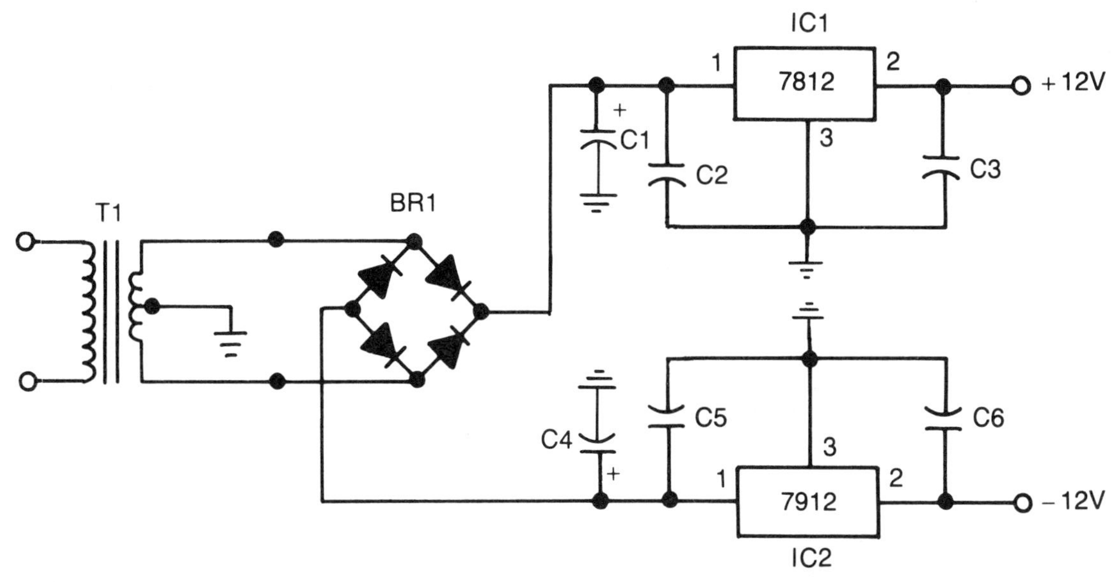

Fig. 8-11. A 12-volt bipolar power supply. Be very careful in designing and constructing this type of supply to observe the orientation of the components with respect to ground, and to provide mechanical and electrical isolation where necessary.

While the positive section uses a regulator from the same 78*xx* family that provided us with regulation for the single-ended supply, the negative section requires a device from the 79*xx* family, which have negative-voltage inputs and outputs.

You have to observe certain precautions in using negative voltage regulators. In a single-ended, positive-voltage, circuit, ground potential is negative with respect to the output (Fig. 8-12*A*). In a negative supply, however, ground is *positive* with respect to the output (Fig. 8-12*B*). And, while the positive and negative sides of the supply can share a common ground, the positive and negative portions of the circuit must be isolated from one another, and the proper component polarities observed.

For example, while it's true that the electrolytic capacitor in the negative half of the supply is connected "backward" (with its positive side going to the side of higher potential—ground) the positive and negative halves of the supply are not simply mirror images of one another. This is because positive and negative voltage regulators have different pinouts and thus require different physical layouts. Both the positive and negative devices have one of their three pins connected to the shell or case. In the case of a positive regulator such as the 7812 shown in Fig. 8-13*A*, this is the ground pin, which makes it very simple to ground the device. You just mount it on the chassis, which is also at ground poten-

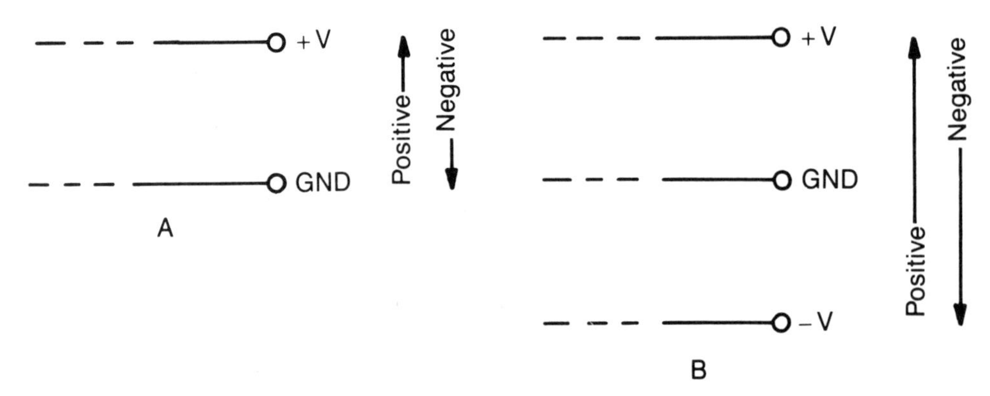

Fig. 8-12. It is important to remember that, unlike the situation in a single-ended (single-output) power supply (*A*), in a bipolar power supply ground potential can be both lower and higher than that of the output voltages (*B*).

tial. You should still also ground the appropriate pin of the regulator if one is provided, as is the case with tab-type cases.

The shell of a negative-output device is *not* its ground, though—it's its *input* (Fig. 8-13*B*)! If you were to connect the shells of both the positive and negative devices to a common point, such as the power-supply chassis you might also be using as a heat sink, you would in fact be connecting the input of the negative regulator to ground! To say the least, this would be disastrous. Therefore the negative regulator must be isolated both electrically and mechanically from the positive one.

You achieve this isolation by ensuring that the case of the negative regulator does not touch the chassis. Sometimes, where a lot of heat-sinking is not required, tab-type devices can be circuit-board mounted, avoiding the chasis-ground-short problem. You can use a clip-on heat sink if you want to, provided it does not come in contact with any other

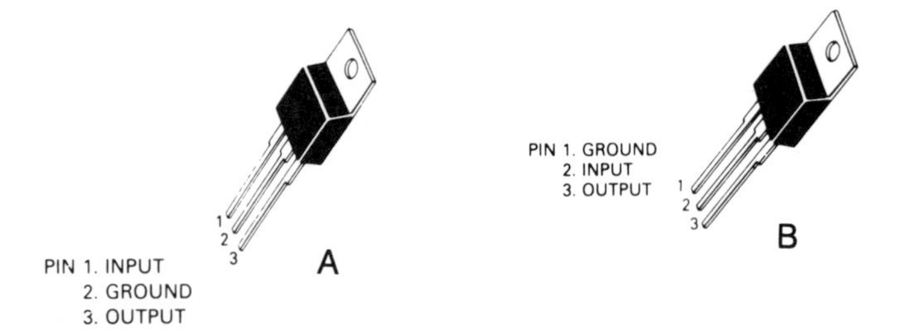

Fig. 8-13. Although they look alike, the pin numbering of positive (*A*) and negative (*B*) voltage regulators is different with regard to function. In addition, while the case of a positive regulator is connected internally to the ground pin, the case of a negative regulator is connected to the input pin!

part of the circuit. Higher-capacity TO-3-type devices though, require chassis mounting. Isolation is provided, as shown in Figs. 8-14 and 8-15, by insulators. These insulators are thin sheets of mica or plastic with oversize holes in them to permit device leads or mounting hardware to pass through.

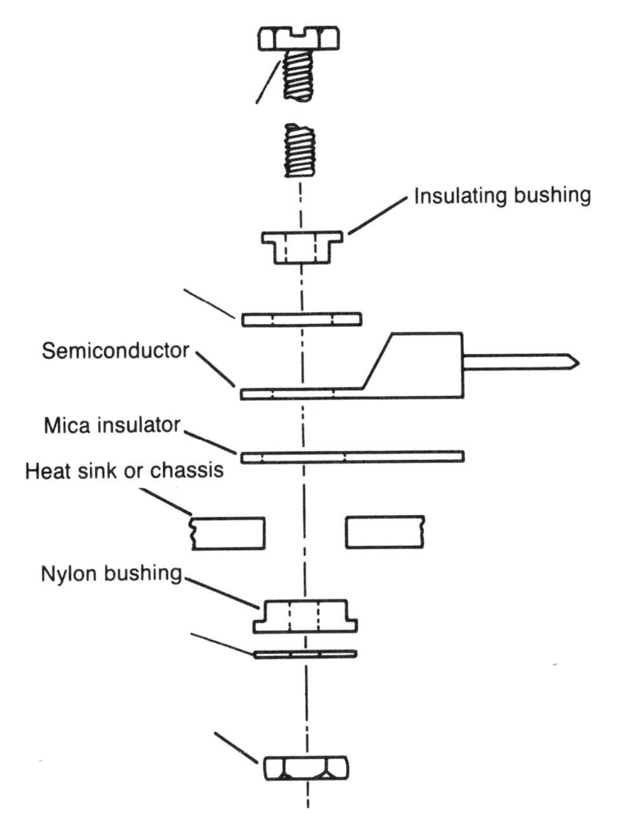

Fig. 8-14. Mounting a TO-220 device on a heat sink. If the tab is intended to be at ground potential, it can usually be connected directly to the chassis without the benefit of a mica insulator and its associated bushings. If you do use the insulator, spread both sides of it with a thin layer of silicone grease to facilitate heat transfer.

While these insulators do a good job of electrical isolation, they also tend to block heat flow between a device and the object—be it chassis or heat sink—to which it is attached. Applying a thin layer of silicone grease to both sides of the thin insulator facilitates heat transfer. Be careful working with this compound—if you get it on your clothes you may have trouble getting it out. Use a cotton swab to remove any excess grease after the device and its insulator have been mounted.

Be careful to mount components such as negative voltage regulators so you cannot inadvertently—as with a stray lead or a screwdriver—ground their cases. In extreme cases you might want to enclose the device so nothing can touch it. And, needless to say, the negative-output terminal of the supply, where you connect it to the outside world, must also be electrically isolated from the chassis.

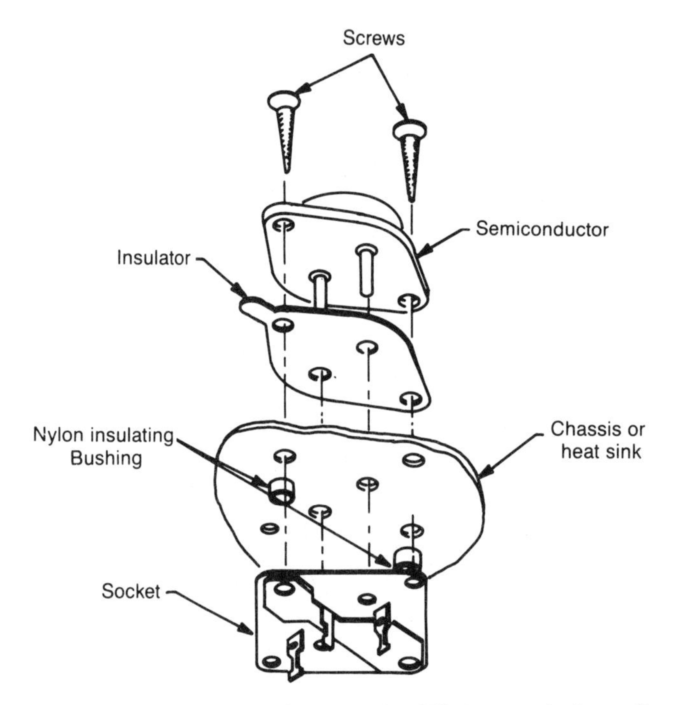

Fig. 8-15. This technique is used for mounting TO-3-case devices. Some TO-3 sockets have insulating bushings molded into them. Copyright Motorola, Inc. Used by permission.

NON-STANDARD VOLTAGES

Positive and negative voltage regulators come in a choice of several values—5-, 12- and 15 volts are widely used, and others are available. But what do you do if you need an output voltage that does not match that of any readily available regulator? You fake it!

The output voltage of a regulator is referenced to ground. Therefore, if you can change the way the device "looks at" ground potential, you can change its output. Figure 8-16 shows a 12-volt regulator connected so as to deliver an output of 13.4 volts, just about the voltage of a fully charged automobile battery. Between the regulator and ground are two diodes connected in series. These diodes create a "pedestal" above ground potential equal in height to the sum of the voltage drops across them—twice 0.7 volts in this case, or 1.4 volts. Added to the twelve volts the regulator thinks it's delivering, this produces an output of 13.4 volts.

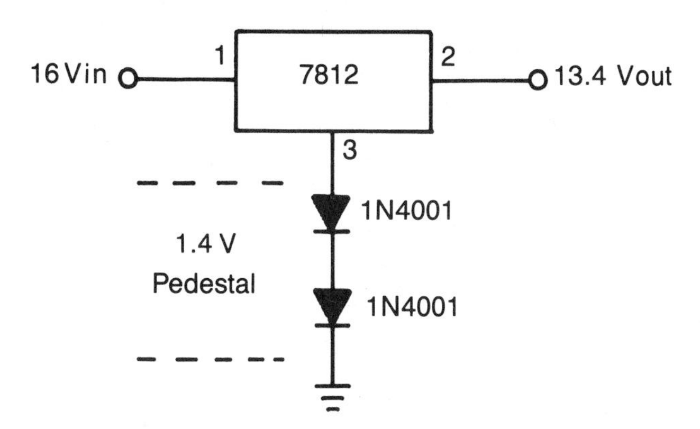

Fig. 8-16. Silicon diodes, with their 0.7-volt voltage drop, can be used—in series if need be—to establish a voltage regulator on a pedestal above ground potential and increase the output of the device accordingly.

Zener diodes can be used in place of the more conventional devices to produce other voltages and you can, of course, use this same trick in negative-output supplies as well. Remember, in this case, though, to reverse the polarities of the diodes.

ADJUSTABLE REGULATORS

While the trick just described will allow you to manipulate the output of a fixed-voltage regulator, it is often much simpler just to use a device that has been specifically designed to have an adjustable output. One such adjustable regulator is the 317, and a circuit that uses it to produce voltages up to about 12 volts is depicted in Fig. 8-17. Resistors R1 and R2 develop a voltage pedestal with respect to ground. As a result, the regulator outputs a voltage determined primarily by R2. Because the device is not referenced directly to ground potential, it must be isolated from ground as described above in the section on negative fixed-voltage regulators.

Both R1 and R2 can be low-power, 1/4-watt components. They are used by the regulator's internal circuitry. Their power-handling capacity has nothing to do with the output current of the regulator. Resistor R1 can be just about any value between 220 and 1000 ohms. In practical terms, the output voltage of this circuit is determined by the formula:

$$V_{OUT} = 1.25 \times \frac{1 + R2}{R1}$$

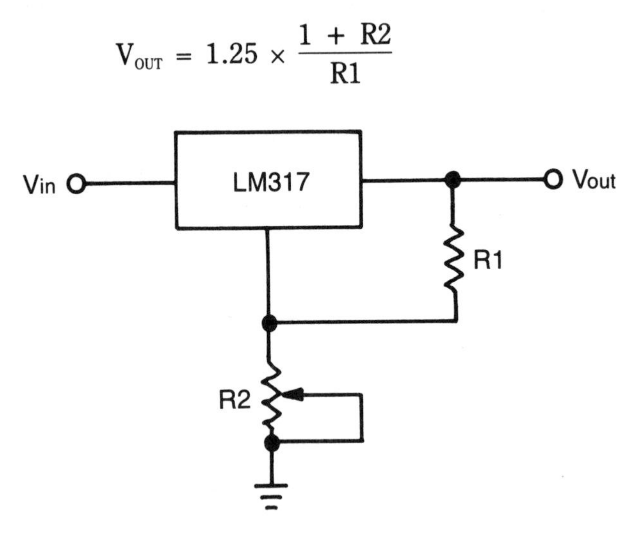

Fig. 8-17. The LM317 is a good example of an adjustable voltage regulator. Its output is established primarily by the value of potentiometer R2.

9
Circuits That Flash

When you stop to think about it, at the heart of lots and lots of circuits are time constants. That is, many circuits do nothing but repeat the same action over and over again at fixed (or sometimes variable) time intervals. Circuits in this class include all kinds of timers, flashers, and audio (and radio-frequency) oscillators. The time constant of these circuits is what determines how frequently the action will occur, or to put it another way, its frequency.

One integrated circuit that puts it all together when it comes to timing circuits is the 555 timer (Fig. 9-1). It is so versatile that entire books have been written describing the various uses to which it can be put. We'll look at the 555 in this chapter, and use it as a basis for examining some of the other components that are used in timing circuits.

555 BASICS

With just a few simple components and a 555 IC, you can construct a multitude of devices. And, once you have an idea of how to construct 555 circuits, you will easily be able to design circuits using other timer ICs as well.

The basic 555 integrated circuit is available in regular and low-power CMOS versions, and can be operated from supply voltages anywhere between about five volts and fifteen volts. It can source (output) or sink (pass to ground) currents up to 200 milliamperes. The 556 is a dual-timer version of the 555, offering two 555 ICs in one fourteen-pin package, and a quad version, the 558, contains four of them in a single

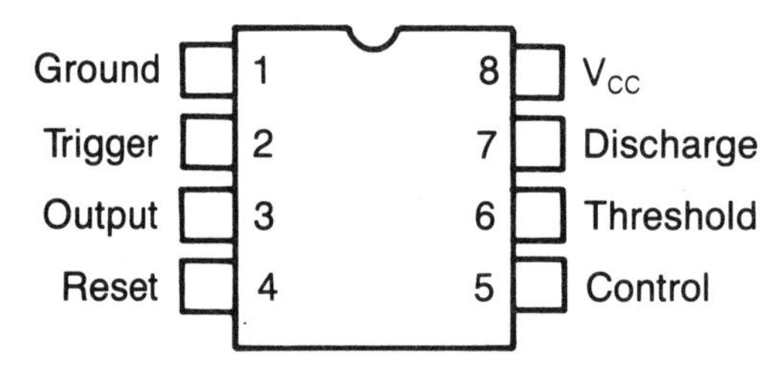

Fig. 9-1. The 555 timer IC is one of the most versatile—and fun to design with—integrated circuits on the market.

eighteen-pin DIP. Since two eight-pin 555s will fit nicely in one 16-pin socket, though, don't go out of your way looking for special devices if you don't really need them.

The 555 has three basic operating modes: timer, monostable (one-shot) multivibrator, and astable multivibrator (oscillator). For the first two types of operation only two components, a resistor and a capacitor, are needed to establish a time constant. Adding a second resistor allows you to use the IC in its third, astable, mode.

Figure 9-2 shows a 555 time delay circuit that will cause an LED to light after a certain interval has passed. The action of the timer is initiated by momentarily grounding capacitor C1 with switch S1 and discharging it to ground. If you had a large enough capacitor you could produce a delay of several hours. The formula for determining the components for a delay is:

$$T = 1.1 \times RC$$

The value of the delay, T, is in seconds, the value of R is in ohms, and the value of C is in farads. Remember to convert a value in microfarads to farads by dividing by a million (or multiplying the value in μF by 0.000001) before trying to solve the equation, otherwise your calculations will be way off!

Since resistors are less expensive than good stable capacitors, it makes sense to start with a high-quality capacitor you have on hand and then make your adjustments by finding—and even breaking down and buying if you have to—a suitable resistor. You can use a variable resistor to get started and once you've found a setting that gives you the delay you want you can measure the resistance of the potentiometer and substitute a fixed resistor of that value.

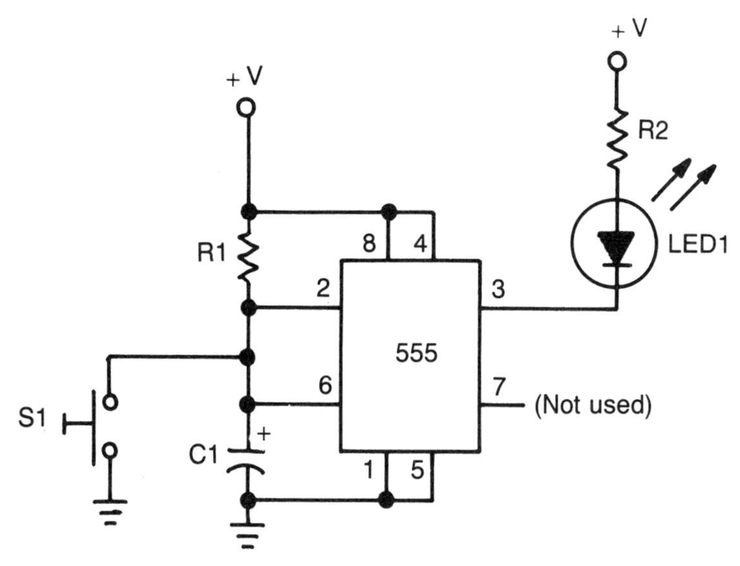

Fig. 9-2. This 555 time delay circuit causes LED to light after a predetermined interval has elapsed. Note how the unused pins of the IC, pins 4 and 5, are tied to +V and ground, respectively. This ensures your knowing their state at all times (since you put them into it).

Let's say that you have a good stable 47μF capacitor on hand to use in a time delay circuit, and that you want a delay of 12 seconds. That means that the formula becomes:

$$12 = 1.1 \times R \times 0.000047$$

Rearranging this equation to balance it so as to solve for R (you do this by doing the same things to both sides of the equation until R is on one side and everything else is on the other), the component you're going to rush out and buy or scrounge around in your junk box for, the equation becomes:

$$R = 12 \div (1.1 \times 0.000047)$$

or

$$R = 12 \div 0.0000517$$

or

$$232{,}108.31 \text{ ohms}$$

which is not what most engineers would regard as an off-the-shelf resistor value. However, it is a simple matter to connect a 220k resistor—which is a common value—in series with a 25k or 50k potentiometer. The two together will allow you to come up with the correct resistance. Indeed, given the fact that the tolerances of the components, particularly the capacitor, can result in their actual values being rather different from their marked values, having such an adjustable-value component in the circuit is a very good idea if you want to match the time constant you calculated precisely.

Once you're in the ballpark, by the way—and a little experimenting will soon give you an almost intuitive sense of what values will give you what timing ranges—you can easily double or halve the time constant by doubling or halving the appropriate component. And, of course, changing the time constant by other factors is just as simple.

If you own or have access to a computer, you may want to write a little BASIC program that will simulate this circuit through the equation above and allow you to "plug in" various component values to see what the time constant will be. Such programs appear from time to time in various electronics or computer magazines, but it's a simple matter to write your own. (And, by doing so, you'll gain a better understanding of how the values of the components in the circuit interreact.)

The maximum resistance you can reliably use in a 555 timer circuit is about 3.3 megohms. Just for the fun of it, let's see what would happen if we used a 3.3 megohm resistor with one of the new half-farad capacitors you can pick up for a couple of dollars:

$$T = 1.1 \times 3300000 \times 0.5$$

or

$$T = 1{,}815{,}000 \text{ seconds}$$

or

30,250 minutes

or

504 hours (21 days)

Sometime when you have a few extra bucks and nothing better to do, why don't you try it? Add a low-value (several ohms) high-wattage resistor in series with switch S1 for this application; half a farad's worth of electrons should not try to go directly to ground all at once. And remember to come back in three weeks to check on the LED!

Other Circuit Features

There's another part of this circuit from which we can learn a couple of things, the output end. When the time delay is up, and the output of the 555 becomes active you may get a surprise. The output does not put out a voltage—instead it goes almost to ground potential and, instead of *sourcing*, or generating, a current, it becomes capable of *sinking*, or passing to ground, a current of several hundred milliamperes. This, perhaps unexpected, switch to ground potential is a rather common thing in the world of semiconductors, and particularly in ICs. It is easier to fabricate components that will sink current than it is components that will source it. In logic circuits, you frequently find what's called "negative-going" logic where an output is recognized by the fact that it has gone to ground potential from either a five-volt level or from a high-impedance state. Sometimes, where one stage of logic triggers another, alternate negative-going and positive-going stages are used to simplify design.

Now, since the output of the timer goes low (to ground potential), the current for the LED has to come from somewhere else. Fortunately we can connect the LED to the same positive voltage that powers the rest of the circuit. The indicator won't draw any current until the active low of the 555 establishes a path to ground for it. (Note, by the way, that the cathode, or lower-potential end, of the LED is the end of the device that is connected to ground. The cathode of any diode—light-emitting or otherwise—must always be negative with respect to its anode if current is to flow.)

LEDs are great gadgets—they're inexpensive, don't consume much power (only about 20 mA), and if you get enough of them together with a few logic ICs you can create some pretty mind-boggling displays. They do have a problem, though. The problem is that, if they are connected between a source of voltage and ground they will gobble up more than is good for them if it is available, and will quickly burn themselves out. They will go out in a blaze of glory, it's true, but then you'll have to replace them, which is not something you want to have to keep doing.

Most LEDs are intended to draw about 20 mA at a potential of 1.3 volts. Under these conditions they will last for many years. Left to themselves, however, they will try to "go for it all," with disastrous results. And that brings us to resistor R2, which is connected in series with the LED.

Resistor R2 is known as a *current-limiting resistor*. As its name implies, its purpose is to limit the current that can flow through the LED. It does this by reducing the voltage applied to the LED to the 1.3 volts at which the device will best operate. To determine the value of the resistor we must know two things: the voltage drop across it and the current through it. We can then use Ohm's law to arrive at its value.

We know from the LED's spec sheet that it should draw 20 mA. Since the LED and the resistor are in series, this is the current that will also flow through the resistor. And we know that the voltage going into the resistor is five volts, and that at the other end, going into the LED, we want 1.3 volts. The voltage drop across the resistor, then, will be $5 - 1.3$, or 3.7 volts. This is very important to remember when calculating the value of any current-limiting resistor—you must use the voltage drop, the difference between the input and output voltages in the Ohm's law formula. Using just one voltage or other will not work—the solution will be incorrect.

Applying the current and voltage values for this case to Ohm's law, we get:

$$R = E \div I$$

or

$$R = 3.7 \div .020$$

or

$$R = 185 \text{ ohms}$$

A 180-ohm resistor (or even a 220-ohm one) will do the job nicely. A $1/4$-watt-unit will allow plenty of margin. (You can check this by calculating the power that has to be dissipated by the resistor—3.7 volts $\times$ 0.020 amperes, or 74 milliwatts). That's a lot less than the 250-milliwatt rating of the resistor and you could even use a $1/8$-watt (125 milliwatts) unit safely.

Remember how to calculate the value of a current-limiting resistor—you'll use one almost every time you use an LED, and there will be plenty of other situations in which you wish to prevent a device from drawing more than a certain amount of current. This is how you do it.

MONOSTABLE OPERATION

In its *monostable* mode the 555 timer becomes a "one-shot" device, a single-pulse generator. When it is triggered, the output at pin three of the IC goes high. The circuit is shown in Fig. 9-3. The values of R1 and C1 determine the duration of the pulse at the output. While this circuit is similar to that of the time-delay one in Fig. 9-2, it is triggered differently, by taking the TRIGGER input briefly to ground. This can be done as shown, with switch S1, or could be the result of the output of another circuit (perhaps another 555) going low and taking the trigger input with it to that level. And, although the RESET pin, pin 4 of the IC, is shown unused and connected to the supply voltage, it can be used to reset the pulse generator by momentarily grounding it.

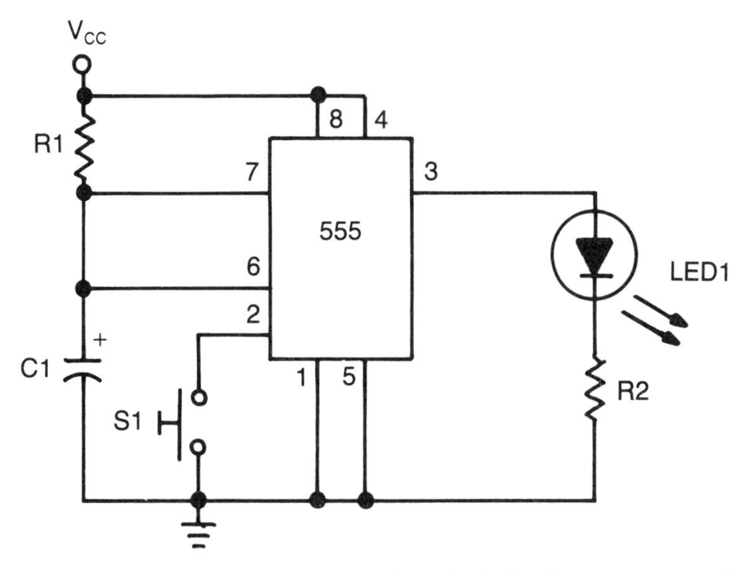

Fig. 9-3. This monostable multivibrator (one-shot) circuit generates a pulse of a specific length when switch S1 is depressed momentarily. The duration of the pulse is determined by the values of R1 and C1.

The same formula, $T = 1.1 \times RC$, that was used to determine the duration of the time delay in our first 555 circuit is used here to calculate the duration of the output pulse. For a pulse lasting a fifth of a second (0.2 second), with a 47k resistor, the equation becomes:

$$T = 1.1 \times RC$$

or

$$0.2 = 1.1 \times 47{,}000 \times C$$

$$C = \frac{0.2}{1.1 \times 47{,}000}$$

$$C = \frac{0.2}{51{,}700}$$

C = 0.00000386 *farads* or 3.86 microfarads

A capacitor with a value of 3.3 or 4.7 microfarads will get you into the ballpark and increasing the value of the resistor will allow you to use a smaller-value, and therefore less expensive, capacitor.

Note that the LED in this circuit, which will illuminate for a fifth of a second to indicate the pulse, is connected oppositely from the one in the previous circuit. This is because the output of the 555 goes high in this application, while it went low in the other. Because the current flow is reversed, the polarity of the device must also be.

Let's return to the RESET pin, pin 4, for a moment. What we do with it illustrates another important practice in electronic design. You might think that if you were not going to use this pin you might be able just to ignore it. Not so! You always want to know what the unused inputs of your devices are doing, even if they're not doing anything. If you take steps to make sure they do nothing, then you will always know exactly what's going on in your circuits. Therefore, it is good design practice to tie unused inputs either high (to the supply voltage) or to ground, whichever one ensures a stable, no-action, state.

Now that you know how to design a timer and a pulse generator, you may want to play around with having the first trigger the second. You can use two 555s, or a 556 dual-circuit package.

ASTABLE CIRCUITS

Now comes the fun part! Just by including a second resistor, you can cause the 555 to operate in the *astable*, or oscillatory, mode. "Astable" means "not stable," and that instability is the basis for oscillation. The circuit goes to one state, can't stay there, and falls out of it into another. However, it doesn't like it there either, so it reverts to the first. And so on. And so on. The choice of timing components determines how frequently the circuit will change from one state to the other and back again, thus setting its frequency.

As was mentioned at the beginning of this chapter, many types of electronic circuits are based on oscillators with various time constants. Slow frequencies—with periods a little bit greater than or less than a second—are good for such things as light flashers. Decreasing the period (which increases the frequency) to several hundredths or thousandths of a second allows you to build oscillators whose output is in the audio-frequency range (20-20,000 Hz), and that can be connected to speakers to produce tones. Decreasing the period even further gets you into the ultrasonic range and, far beyond that, into the area of radio-frequencies. The 555 can be made to change states at a rate of as much as a megahertz—1,000,000 times a second—or so. This frequency is in the middle of the AM radio broadcast band!

Figure 9-4 shows a 555 connected for astable operation, a simple "blinky" circuit that will flash an LED. This circuit works on the principle that capacitor C1 charges through resistors R1 and R2, and discharges through resistor R2. This discharge occurs when the charge on the capacitor reaches a level of about 2/3 its capacity. Remember this number— 2/3, or 0.666—you'll use it often in timing circuits.

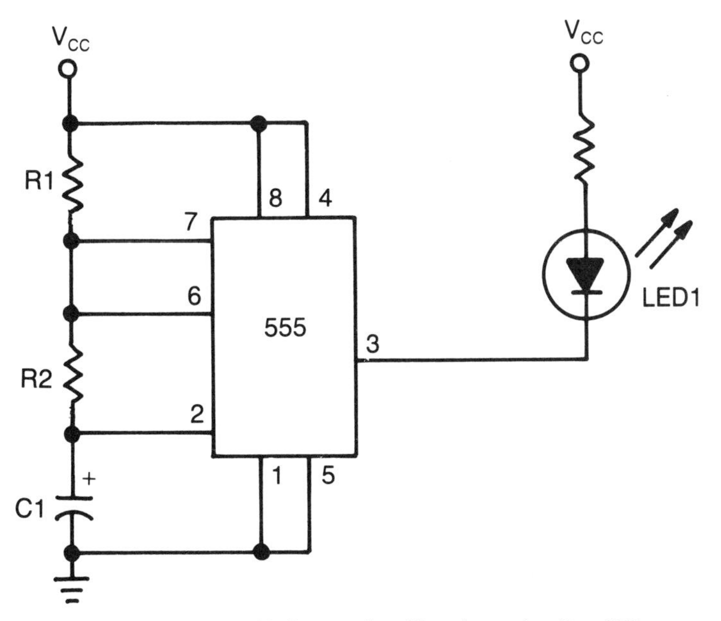

Fig. 9-4. In its astable multivibrator (oscillator) mode, the 555 can generate square waves whose frequencies have periods ranging from minutes to microseconds.

The values of the components determine how frequently the LED in the circuit will flash and the duration of each flash, as well as the time that elapses between flashes. The ratio of flash on-time to flash off-time is called the *duty cycle* of the circuit; understanding duty cycles can help you to design power-conservation circuits. We'll get to them shortly.

The formula used to determine the values of the three timing components in an astable 555 circuit, C1, R1, and R2, is:

$$f = \frac{1.49}{(R1 + 2R2) + C1}$$

The letter f, of course, stands for "frequency."

According to the application notes for this IC, if R2 is large (say, several hundred thousand ohms) with regard to R1, a duty cycle of very nearly 50 percent, where the LED is on and off for equal periods of time, will be attained. While R1 should not be smaller than 1k, R2 can have values as great as several megohms. The value of the capacitor, C2, can range anywhere from 500 pF to several hundred microfarads—the larger the capacitance, the longer will be the period of the timing circuit, meaning the more slowly the LED will flash. (You may have noted that in general, larger capacitances mean lower frequencies.)

This is a very forgiving circuit—as long as you make R2 considerably larger than R1 something will happen. Because you want to be able to observe the LED flashing, use a large value for C1 and a large value for R2—making the capacitor charge through a larger resistance also increases the period of the timing circuit.

How fast does the LED in the circuit in Fig. 9-4 flash?

$$f = \frac{1.49}{(R1 + 2R2) \times C1}$$

$$f = \frac{1.49}{[1000 + (2 \times 390{,}000)] \times 0.000001}$$

$$f = \frac{1.49}{781{,}000 \times 0.000001}$$

$$f = 1.9 \text{ Hz}$$

Again, if you have access to a computer, you can write a BASIC program that will allow you to input the values of the various components involved and see what the outcome will be. One of the things you'll discover is that the value of R1 has very little effect on the frequency. Its purpose is primarily to establish the duty cycle of the output waveform. And now we can discuss the concept of duty cycle.

Duty Cycle

When you think of a digital waveform you usually conceive of it as looking something like the square wave in Fig. 9-5. In this figure, the high and low portions of the wave are the same length; that is, they have the same duration. This represents a 50% duty cycle—the waveform is high for half (50%) of the time, and low for the other half. This does not always have to be the case, though.

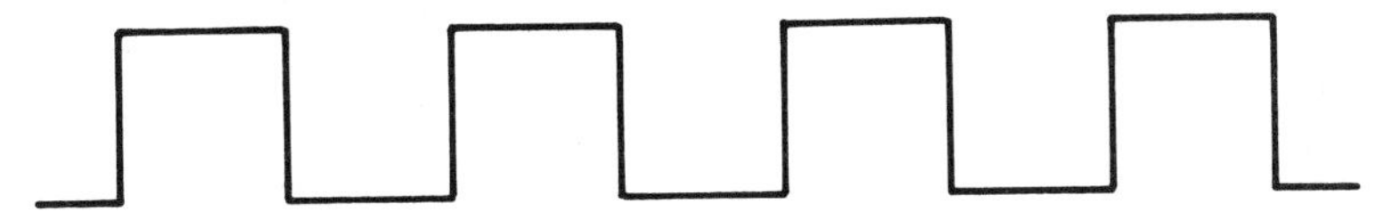

Fig. 9-5. The common conception of a square wave has it spending equal amounts of time at high and low potentials.

The duty cycle of a 555 astable circuit is calculated from the equation:

$$D = \frac{R2}{R1 + 2R2}$$

and the on- and off-times (the time the waveform spends high and the time it spends low) by the formulas:

$$T_{HIGH} = 0.67 \times R1 \times C$$
$$T_{LOW} = 0.67 \times R2 \times C$$

If we make R1 so small that it almost disappears with respect to R2, we can just about disregard it and the formula for the duty cycle becomes simply:

$$D = R2 \div 2R2$$

or 0.5 (50%). In the pair of equations above the only thing that changes from the one to the other is the value of the resistors—the value of C is the same in both, as is the number 0.67. (The value "0.67," by the way, is our old friend representing 2/3 (0.666) of the full charge the capacitor can hold.) Therefore, the ratio of high time to low time is affected simply by the ratio of R1 to R2.

(Use the circuit shown in Fig. 9-6, which adds two diodes to the one in Fig. 9-4, if you need a duty cycle of less than 50%. The diodes isolate the charge and discharge paths of capacitor C1. This circuit will allow you to establish duty cycles from less than 5% to greater than 95%.)

After all this, you're probably wondering what useful purpose, if any, being able to manipulate the duty cycle serves. Let's assume, for the moment that you've established a circuit that flashes the LED at the rate of 60 Hz (the same rate as household ac "flashes" a fluorescent lamp). When it's flashing at a rate this fast it seems to be on constantly; you don't notice the fact that it is actually turning on and off sixty times a second. Suppose, now, that you want the LED to appear dimmer, but still flash at the same rate. You could increase the value of current limiting resistor R3 to limit current flow even further and permit the LED to illuminate less brightly, but that would be wasteful—the power not consumed by the LED would be dissipated as heat by the resistor.

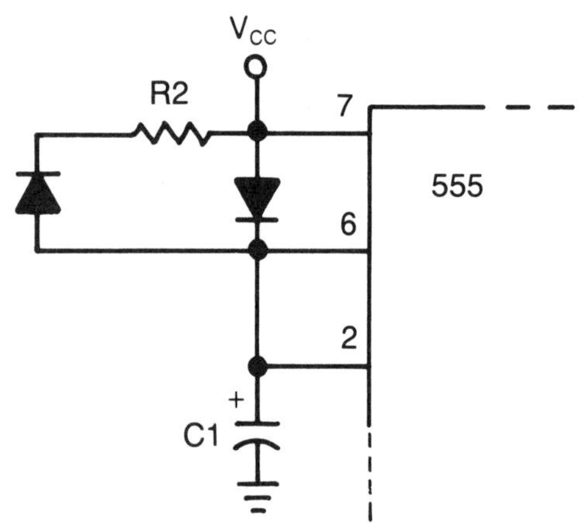

Fig. 9-6. By adding two diodes to the astable circuit shown in Fig. 9-4, the duty cycle of the square wave output at pin 3 of the IC can be varied from less than 5% to greater than 95%.

The more efficient way would be to change the ratio of the indicator's on-time to off-time—in other words, its duty cycle. If it were off more than it were on, it would appear dimmer and, still operating at 60 Hz, no flicker would be noticeable. By decreasing the time the LED was on, you might even save a little power.

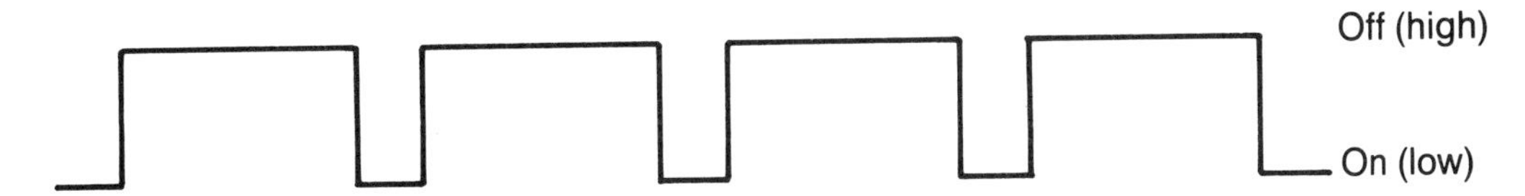

Fig. 9-7. By modifying the duty cycle of the astable circuit so the output is high much longer than it is low, the on-time—and thus the brightness—of an LED connected to the output can be reduced, as can the power it consumes. This technique is much more efficient (and elegant) than the brute-force one of adding a resistor to "burn off" unwanted current.

The LED in the circuit in Fig. 9-4 illuminates when the output of the 555 goes *low*. That means that if you look at the duty cycle "upside down," so the time high becomes the off-time, and the time low becomes the on-time (Fig. 9-7) you can easily adjust it to cause the LED to be on less than half the time and thus appear dimmer. (The LED's actual brightness at any instant is actually the same as it would be with a 50% duty cycle. Since the device is on for less time, though, and off for more, it generates less light and appears less bright.)

This trick of adjusting brightness through varying the duty cycle is frequently used, especially where saving power is an important consideration.

Rearranging the Equation

As we've seen, the formula for finding the frequency of an astable multivibrator built from a 555 timer IC is:

$$f = \frac{1.49}{(R1 + 2R2) \times C1}$$

You still have to plug in the values of R1, R2, and C1, but with a little experience you can easily come up with ballpark figures for two of them and use those to determine the third. What do you do, though, if you know the frequency you want to output, and you have a certain value of capacitor on hand, but need to know what value R2 should be? (As we've seen, the value of R1 is nothing to worry about as long as it's small with respect to that of R2.) What you do is rearrange the equation but, since this is not a book on algebraic problem solving, we'll present you with the equations and then get on to more important things:

$$C1 = \frac{1.49}{(R1 + 2R2) \times f}$$

$$R1 = \frac{1.49}{C1f} - 2R2 \qquad R2 = \frac{\frac{1.49}{C1f} - R1}{2}$$

Modifying the Astable Circuit

Because we've been working with an LED as an output indicator, we've had to keep the frequency of the 555 circuit in Fig. 9-4 low so we could see the LED flash. However, since the 555 can operate at frequencies of several hundred kilohertz, let's see what happens if we increase the output frequency of our circuit just a little.

The circuit, when we last used it, had an output with a frequency of about 2 Hz. If we were to decrease the value of C1 by a factor of 100 by using a 0.01μF capacitor instead of one with a capacitance of 1μF we'd have an output frequency of 200 Hz; and if we used a 0.001μF part the frequency would increase to 2000 Hz. (Remember, changing the value of resistor R2, or capacitor C1, also changes the output frequency of the circuit by the same factor. You can double the resistance or halve the capacitance; the results will be just about equal.)

Those frequencies—200 Hz and 2 kHz—both lie within the audio spectrum, the range of frequencies that we perceive as sound. So, if we connect the output of the circuit to a speaker, we'll be able to hear it as a tone. While you can connect pin three of the IC, the output pin, directly to the speaker, the addition of a capacitor to the path (Fig. 9-8) will smooth out the roughly square-wave waveform and make it sound better. The capacitor should have a value of at least 10μF; values smaller than that may attenuate the higher frequencies output by the oscillator.

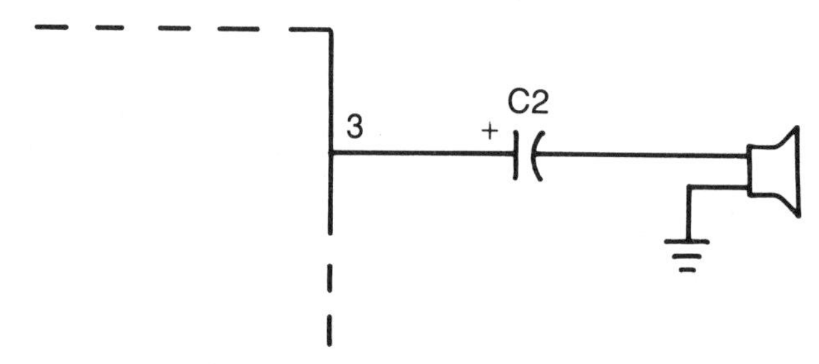

Fig. 9-8. A coupling capacitor with a value of at least 10μF between the output of the IC and a speaker will smooth out the oscillator's square-wave output.

10
Transistors

For about fifteen years, from the late 1950s until the mid-1970s, transistors dominated electronics. In comparison with their predecessors, vacuum tubes, they were much, much smaller and lighter; were enormously more rugged; used hardly any power; and generated much less waste heat—so little as to be almost negligible in most cases. For most purposes they were an engineer's dream come true . . . until the advent of integrated circuits.

Integrated circuits, which consist of a number of transistors, capacitors, and resistors all fabricated together in miniature on the same tiny chip of silicon, came to be electronics' glamour components in the 1970s, and still are. They have diversified into several families and within each family the selection of types is enormous. Because ICs are complete circuits, many of them can be substituted in electronic designs for the equivalent circuits made up of discrete components.

There are ICs for almost every general purpose application you can imagine, as well as a number of rather more specialized ones. As you have begun to see, these circuits can be easily customized to meet specific requirements. What do you need: a timer circuit, a five-volt voltage regulator circuit, a two-watt audio amplifier? Why design (and laboriously construct) a circuit that will convert the binary numbers representing a voltage to signals that light the appropriate segments of an LED display (and display the value of that voltage in the form of numerals) when you can buy one for a few dollars? How about a complete TV receiver on a set of two or three ICs?

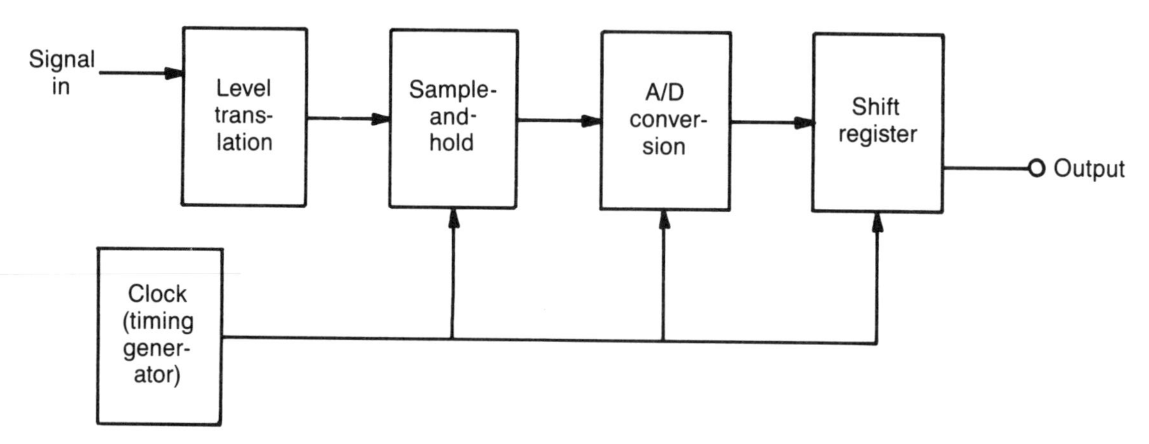

Fig. 10-1. With the variety of integrated circuits, both linear and digital, available today, there is often little design work involved in going from a block diagram to finished device. What there is is mainly a matter of selecting the appropriate building blocks.

For many applications, electronic design has become a matter of drawing a block diagram of a circuit, indicating the function of each portion of the circuit (Fig. 10-1), opening a data book to find the part numbers of the ICs that will perform those functions, and then connecting the ICs one to the other.

ICs are not the answer to every circuit design situation, though. Sometimes you don't need all the frills that can be built into the device that performs the nearest equivalent of the function you need; sometimes it is simpler and less expensive just to knock together a circuit consisting of one transistor and a couple of resistors. This method is less glamorous than the one using an integrated circuit, but it can be a lot more elegant. If all you need is a simple amplifier, an oscillator, or maybe a solid-state switch, a one- or two-transistor circuit could be the best choice.

A ONE-TRANSISTOR AMPLIFIER

The little 555-timer audio oscillator from the last chapter can source or sink a surprising amount of power—almost one watt with a five-volt supply—without getting too hot. Sometimes, though, you may want an amplifier to drive a bigger or less efficient speaker, or a circuit with just a bit more control—say, one that allows you to control its output level conveniently. A one-transistor amplifier might provide you with what you need. Such an amplifier is illustrated in Fig. 10-2.

This circuit uses a 2N3904 transistor, a general-purpose NPN type, although, of course, just about any general purpose NPN transistor will

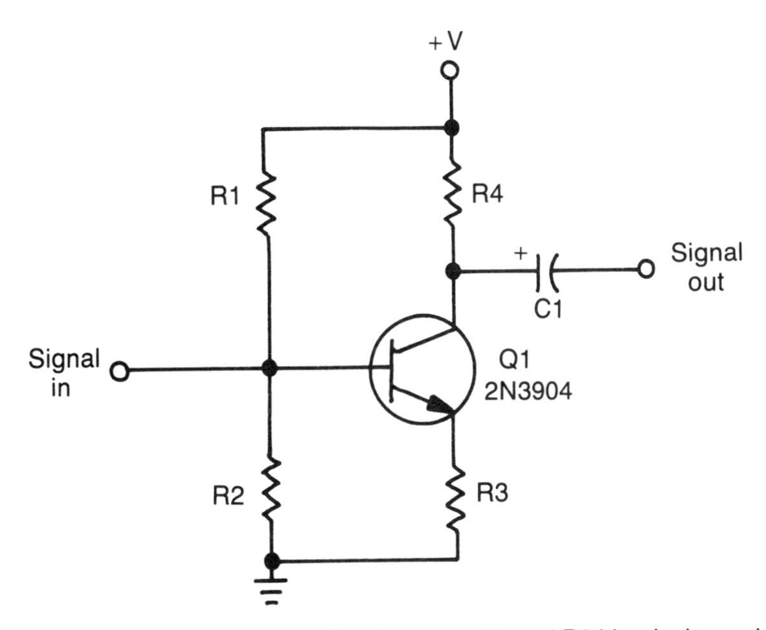

Fig. 10-2. A simple transistor amplifier. Resistors R1 and R2 bias the base about halfway between +V and ground. By reversing the +V and ground connections, a PNP device could be used in place of an NPN one.

do. Whether you use an NPN or PNP type device in your work depends primarily on what voltages you have available in the circuit. Since a transistor must be forward biased to accomplish anything, the collector of an NPN device must be connected to a positive voltage source; that of a PNP transistor would be connected to a negative source. Generally you design analog circuits so that voltages get more and more positive as you progress through them. Therefore, NPN devices are frequently the devices of choice.

Many NPN transistors have a *complementary* PNP device. Complementary devices are those that are essentially similar but are of opposite polarity—positive and negative charges, for example, might be considered complementary charges. Complementary transistor pairs such as the NPN 2N3904 and its complement, the PNP 2N3906, are useful in circuits where devices of similar characteristics but of opposite polarity are required. It is useful to be able to use both types in something like an ac circuit, where one device can handle the positive-going half of the waveform and the other can handle the negative-going half. Some types of amplifiers (push-pull amplifiers, for instance) divide the labor exactly this way and complementary transistor pairs are a necessity.

Because of the way they are constructed, NPN transistors have a slight performance edge over their complementary PNP brethren. They can generally operate at slightly higher voltages and currents, have better gain under similar circumstances, and have somewhat higher maximum operating frequencies. This is not to say that NPN devices are intrinsically better than PNP ones; it's just that that's the way things are.

BIASING

In Chapter 5 you saw that there appear to be more transistor specifications than anyone would know what to do with. You also saw that most of these can be ignored unless you are exceedingly demanding in your requirements. There are a couple of specs, though, that it is very useful to know in designing simple transistor circuits. The most important of these is current gain (sometimes called "small-signal current gain" and also as "beta"), symbolized by the notation "h_{FE}." It is also necessary to consider the current that will flow through the base-collector junction and to provide a resistor to prevent its getting out of hand and destroying the device. The specification to check is "I_C," for "continuous collector current."

The four resistors in the circuit depicted in Fig. 10-2 are necessary to control the current flowing in the transistor. You might occasionally find circuits that use fewer than this number, but these four are what you need to build a reliable transistor amplifier.

Transistors are current-operated devices and it is important to remember this. It is often tempting to think of what happens in solid-state circuits in terms of voltages—indeed, sometimes it's necessary, but in general the values of the components used in transistor circuits are determined by the currents, not the voltages, that will be developed across them. And, if a voltage is involved, there is, of course, always a current to be considered as well. While there is a voltage gain in the type of amplifier circuit used in Fig. 10-2, there is also a current gain and it is primarily this that will affect your calculations.

Depending on how they are used, transistors can provide current gain, voltage gain, or a combination of the two which, as you might have anticipated, is called power gain. The amplifier in Fig. 10-2 provides this combination.

Just as it is necessary to limit the current that can flow through devices such as LEDs, so it is with transistors. This, when you stop to think about it, is not surprising since both are semiconductor devices that are not too distantly related and that have a number of characteristics in common. As with LEDs and diodes, there is a certain voltage drop to contend with—0.7 volt in the case of silicon transistors such as the 2N3904, the same as you find across a silicon diode. Again, not a surprising fact. To overcome this voltage drop it is necessary to forward bias the transistor. This bias voltage is established by resistors R1 and R2. For all intents and purposes, their values can be equal, since the potential developed across R1 should be about 0.7-volt higher than that across R2, the value of the first should be a bit larger than that of the second.

These resistors do two things. First, they allow a forward bias to be developed across the emitter-base junction of the transistor, allowing current to flow. Second, the resistors form a voltage divider between the supply voltage and ground and, since the voltage drop across each resistor is the same, the voltage at the midpoint is half the supply voltage. This nicely sets the base of the transistor at a potential that will allow it to pass signals both above and below an artificial ground potential—in other words, the full wave of an ac signal (Fig. 10-3). If this were not done, the input waveform would be clipped and distorted during amplification.

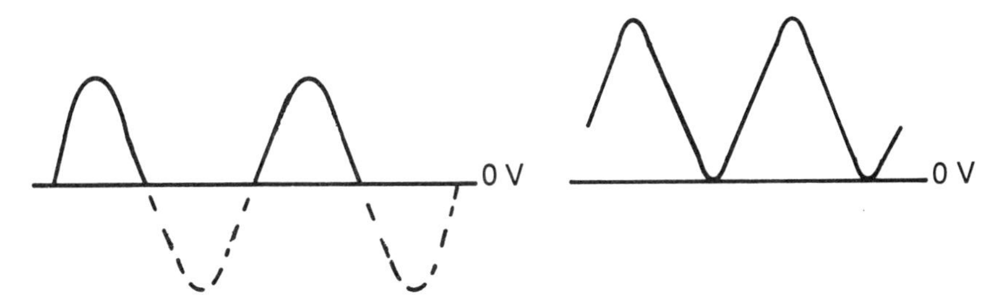

Fig. 10-3. Proper biasing prevents the loss of part of a waveform, avoiding a serious cause of distortion. Reverse biasing (forcing the bias of a device into the negative region) can be used to induce distortion, or to selectively control signal flow.

The value of resistor R1 can, as a rule of thumb, be made about five times the impedance of the transistor. For devices such as a 2N3904 this would work out to about 22k. The other resistor in the voltage divider, R2, should be somewhat smaller than R1, perhaps about half its value.

Resistor R3, between the emitter of the transistor and ground, sets a current flowing through the device by establishing a voltage drop between it and ground. Returning to the water-electricity analogy, you can think of this as a waterfall. If the stream were perfectly level, it would be a still pool and there would be no flow. The resistor establishes a difference in height (electrical potential), and because of this difference a current can flow. Typically, the value of an emitter resistor is 470 ohms. It can be a little higher, but should not exceed 1 or 2k.

Finally, the value of R4 is used to set the voltage gain of the amplifier (the current gain, remember, is established by the transistor's h_{FE}). This gain is equal to the ratio between R4 and the emitter resistor, R3. A tenfold voltage gain is a good general-purpose target; therefore R4 should be about ten times the value of R3. If R3 is 470 ohms, R4 can be 4.7k or 5.6k or thereabouts. Since power is the product of voltage and current, this amplifier has a power gain of approximately 2000—the product (10×200) of the voltage gain and the h_{FE}.

Sometimes in ac amplifiers a capacitor is added in parallel to emitter resistor R3 (Fig. 10-4). This capacitor, which in an audio amplifier should have a value of at least 10 microfarads, prevents R3 from degenerating the performance of the amplifier. Without the capacitor, the gain of the amplifier is simply the ratio of R4 to R3. With it, it is substantially greater.

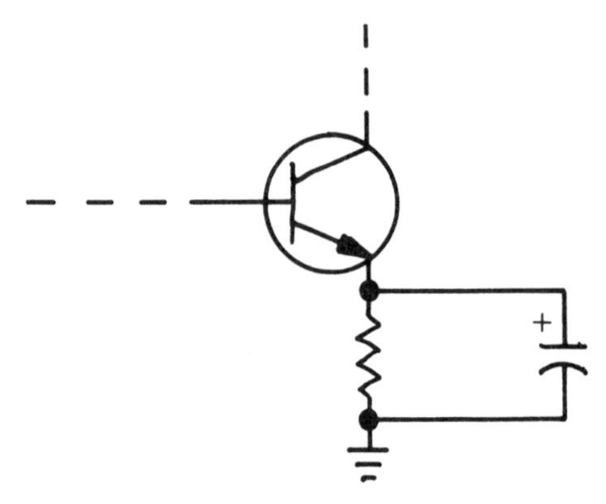

Fig. 10-4. Adding a capacitor in parallel with the emitter-bias resistor reduces the gain of the amplifier, but improves the circuit's ac performance.

The last component in this circuit to be considered is capacitor C1, which couples the output of the amplifier to a speaker. This capacitor's primary purpose is to prevent the supply voltage, which is present on the collector, from passing on to the speaker and through it to ground. Depending on its value, C2 can also affect the ultimate frequency response of the amplifier. Values smaller than about 10 microfarads will cause the lower frequencies to be attenuated. A 10-, 22- or even 47 μF capacitor will work well here.

Note that while the cathode (−) side of an electrolytic capacitor is frequently shown facing down or toward the left—that is, toward the lower potential—in Fig. 10-2 it faces to the right. This is because the higher-potential side is to the left, where the capacitor faces the transistor's collector voltage. To the right is only ground (through the speaker winding).

A TWO-STAGE AMPLIFIER

Figure 10-5 shows a two-transistor amplifier. In the previous circuit the output level was fixed as a function of the gain. In this one, gain is added through the use of an additional stage of amplification and, in addition, a means is provided of controlling the output level of the circuit.

Each stage uses a 2N3904 transistor, and the first stage is identical to the circuit in Fig. 10-2. So, for that matter, is the second stage. This amplifier multiplies the gain of the first stage by the gain of the second.

There are two additional components in this circuit that need some explanation. The first is resistor R9, which is actually a potentiometer—

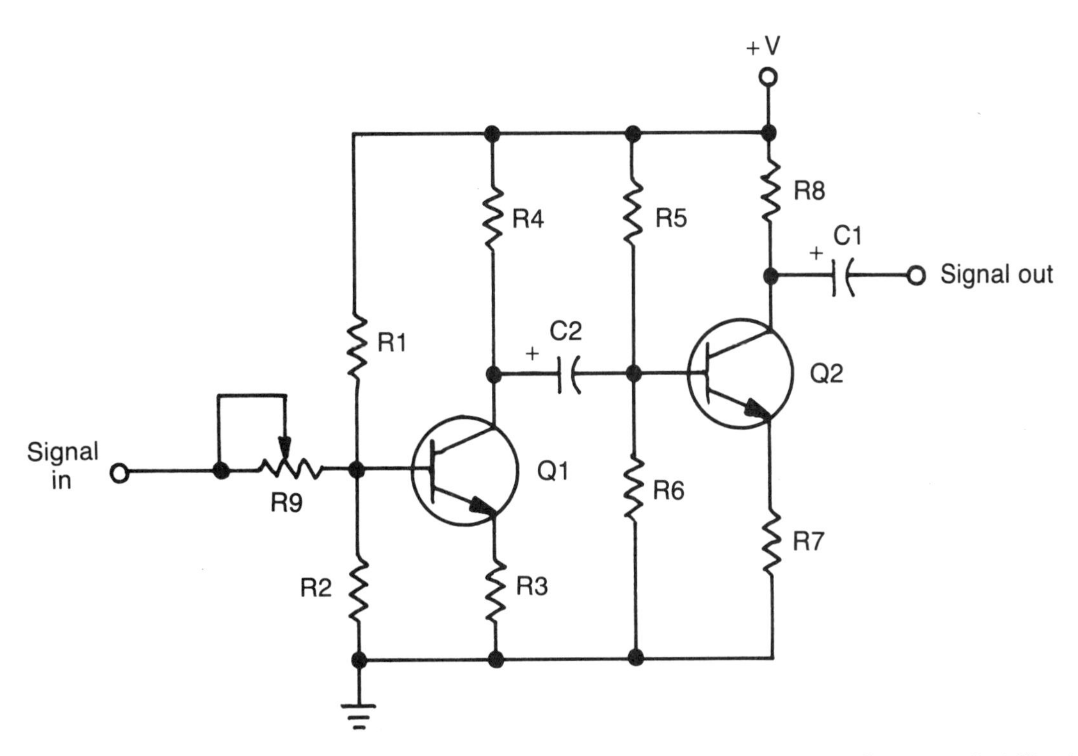

Fig. 10-5. A two-stage transistor amplifier. Note that interstage coupling capacitor C2 is installed "backwards." This is because the collector voltage on Q1 is higher than the voltage applied to the base of Q2.

a variable resistor. It is used to control the degree of amplification, as it were, by controlling the amount of signal applied to the base of transistor Q1. The more signal, the greater the output. As is shown here, you normally want to locate controls such as this as early in the circuit as possible. That way you do not have to manage high currents or voltages, since the initial signal levels are usually pretty low; nor do you incur the large losses in the form of heat that would occur if you were to use a resistor in the latter stages of an amplifier circuit. It is wasteful (and pretty foolish) to go to all the trouble of amplifying a signal and then have to reduce its amplitude after you've done all that work. Much smarter to make all your adjustments early on, where they will not involve as much brute force or waste.

The other component of interest is capacitor C2, between the stages. (In schematics parts numbers, like voltages, generally increase as you go from left to right and from top to bottom. We've bent that rule here to keep the part numbers in the two circuits as nearly identical as possible.) This part is the coupling capacitor, and it serves the same purpose as capacitor C1 at the output of the amplifier in Fig. 10-2. That is, it prevents the collector voltage of the first stage from getting into the next. If this voltage were to be allowed to pass from stage to stage, that from one stage would affect the bias voltage on the base of the next. You usually don't want this to happen, so you prevent it from doing so by placing a capacitor in the path between one stage and the next. Since capacitors pass ac but block dc, the signal gets through but the collector voltage from the previous stage does not. In audio circuits, a coupling capacitor value of ten microfarads or more will do well.

UNITY GAIN

Sometimes a transistor is included between two portions of a circuit for purposes other than power amplification. For instance, you may have a weak signal source that you want to input to several devices, with no change in level. By itself, the signal source cannot deliver enough current for all the loads. What you need is just more current, but not more voltage. A transistor with *unity gain* acting as a *buffer* stage can provide that drive. The term "unity gain" simply means that the transistor amplifies its input voltage by a factor of one, or unity.

Of course, multiplying something by one leaves its value the same. The output voltage is the same as the input voltage, but the current available at this voltage is greater than before. This type of circuit is also known as a *voltage follower*.

A buffer circuit is simple, and one is illustrated in Fig. 10-6. In some respects it resembles the power amplifier we described earlier. In fact, the values of the three resistors in this circuit are determined the same was as they were for the amplifier. Resistor R3, the emitter resistor, is about 470 ohms, R1 is 47k, and R2 is 22k.

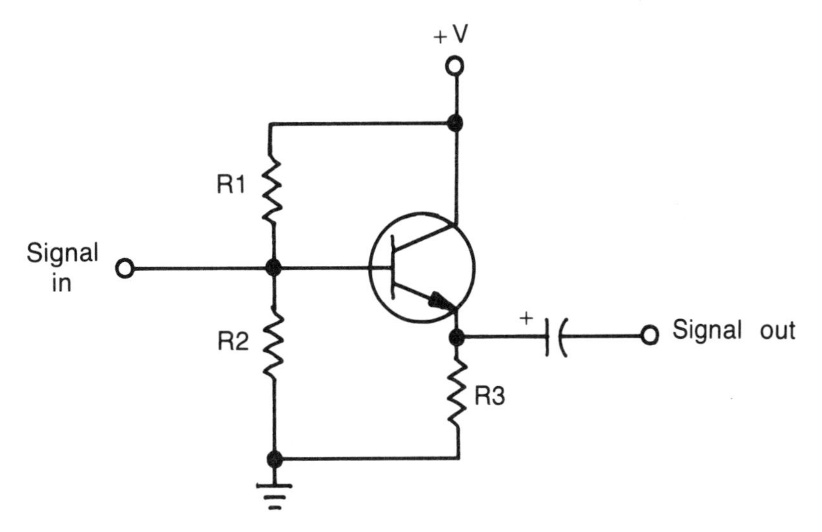

Fig. 10-6. This type of amplifier is known as a buffer. It amplifies current but not voltage, and is used to increase the amount of drive available from a previous circuit stage.

Note that the output of this type of circuit is through the emitter, while the output of a power amplifier is through the collector. Current through the collector is limited by the resistance of the circuitry that is driven by this buffer. The collector of the transistor is connected directly to the supply voltage; there is no need to include a current limiting resistor.

Buffer stages can also be used to isolate one stage from the effects of another, and for impedance matching. Their inputs are high impedance, and their outputs low. This configuration is an example of a common-collector circuit.

A TRANSISTOR SWITCH

One final transistor circuit, perhaps the simplest of all, turns the transistor into a switch. Since a transistor will not conduct until it is forward biased, if it *isn't* forward biased, or if it is reverse-biased, it won't conduct. And if we apply a lot of bias, it will. It's as simple as that. By

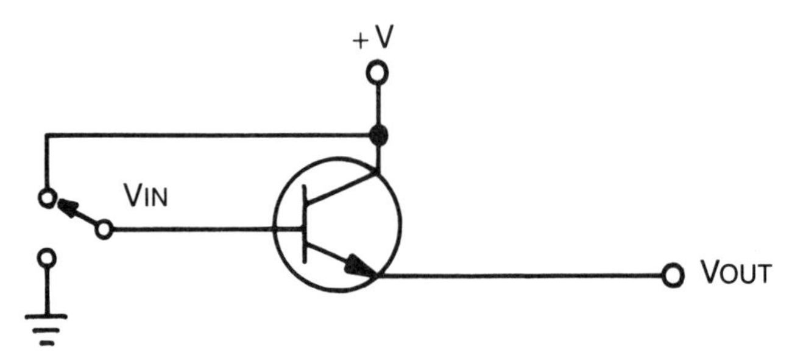

Fig. 10-7. In saturated logic, no amplification takes place. The transistor simply acts as a switch that is on or off depending on whether it is forward biased or not.

connecting the base either to ground or to the full collector voltage we can make its output either zero volts or that voltage. That voltage can then be used to drive other devices, to turn them on.

This forward-biasing of the transistor is called *saturating* it (just as when you drop a sponge into a bucket of water, it becomes saturated with water). A precise forward-bias voltage is not required here, nor is it necessary to use any extra components for stabilization or for setting up a voltage divider. It's enough simply to make the transistor conduct or not, and its output swings between ground potential and the full supply voltage.

Saturated logic is at the heart of most logic ICs, the kind whose function is simply to provide a logic high (a positive voltage) or a logic low (ground) at the output. Saturated logic ICs such as microprocessors can contain thousands upon thousands of transistors working as saturated devices.

11
Operational Amplifiers

The class of linear integrated circuit known as the operational amplifier, or op amp, was discussed in Chapter 6. This chapter explains some of the uses to which these versatile ICs can be put.

The ideal op amp would have characteristics that included:

- Infinite gain
- Infinite input impedance
- Infinite bandwidth
- Zero output impedance

Real-life op amps fall somewhat short of these idealized specifications, but do manage to come about as close to them as you would want them to for most practical purposes. The high input impedance and high gain obtainable from the op amps available to experimenters from any source make them extremely useful devices.

Many of today's op amps are "internally compensated." This refers to the fact that built into them are small—about 30 pF—capacitors that keep them stable and prevent them from going into self-oscillation, something that becomes an increasingly greater possibility as the degree of gain is increased. Uncompensated op amps usually have a wider bandwidth, or frequency response, than compensated devices. They also offer a higher slew rate—that is, their outputs can respond more quickly to changes in input signals than can compensated devices. However, until you get into building circuits that definitely require these extended characteristics, you will get along perfectly well, and more simply, with the compensated variety.

Most op amps require a bipolar power supply, one than can provide positive and negative voltages of equal potential. As long as they are equal, op amp supply voltages can range from about ±5 volts to ±18 volts. Op amp circuits are one place where battery operation is easily possible. A ±9-volt supply can be constructed by connecting two nine-volt "transistor" batteries in series as shown in Fig. 11-1 and using their common connection as a common ground.

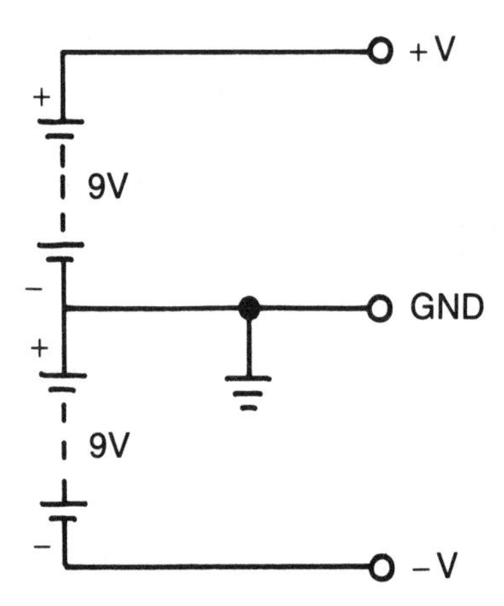

Fig. 11-1. Two nine-volt batteries can be connected in this fashion to provide a portable bipolar power supply for op-amp circuits.

Like most op amps, the 741 (Fig. 11-2), a favorite with experimenters, has two inputs, one marked with a "+" and one with a "–." The one with the plus sign is called the *non-inverting* input, and the one with the minus sign the *inverting* input. Signals fed to the non-inverting input appear at the output unchanged with respect to their polarities—if they were positive to start with, they are positive coming out. However, signals applied to the inverting input get flipped "upside down" (Fig. 11-3). What was positive becomes negative, and what was negative becomes positive. These two inputs of opposite polarities are one thing that make op amps so useful. Signal input voltages can vary over quite a range, as long as they do not exceed the limits set by the supply voltages.

Op amps have two inputs but only one output—the input signals are mixed internally and the result, after amplification or other processing, appears at the output. And, since one of the inputs inverts the input sig-

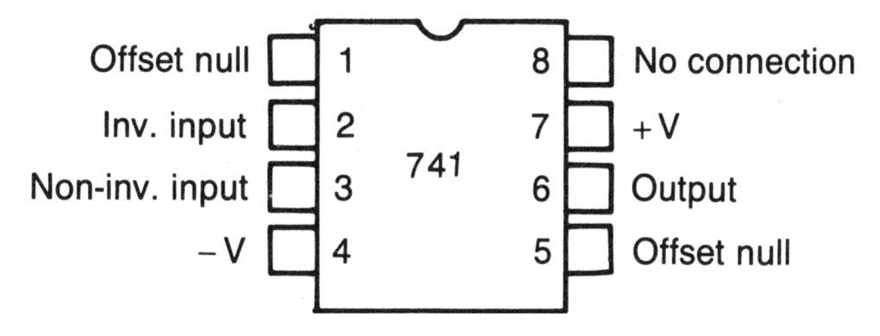

Fig. 11-2. The 741 op amp is another of those "universal" ICs with innumerable uses.

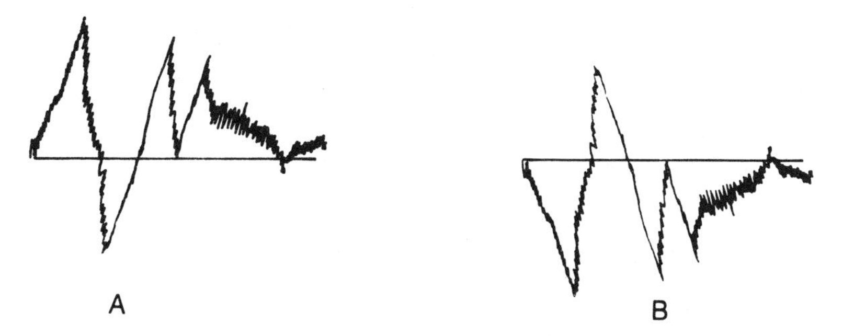

Fig. 11-3. The signal at the inverting input of an op amp (*A*) is the upside-down mirror image of the one at the non-inverting input (*B*).

nal, we can do a number of interesting things with op amps. Figure 11-4 shows the result of applying the same signal to both the "*A*" (inverting) and "*B*" (non-inverting) inputs—namely, *nothing!* That is, because one signal becomes the exact opposite of the other, the two cancel each other within the op amp and there is no output at all. While this may seem a useless curiosity at first, this signal-canceling capability makes possible an entire class of devices known as *differential amplifiers*.

In an "ideal" op amp, the following conditions would be true:

- No current would flow into or out of the inputs.
- Applying a feedback signal from the output to an input of the opposite polarity would result in a differential input voltage of zero.

and even in real-life op amps, these two characteristics are pretty much present.

A differential amplifier gets its name from the fact that it amplifies the difference between two signals. Why would you want to do that? Well, suppose you wanted to amplify a radio signal with a lot of hissy

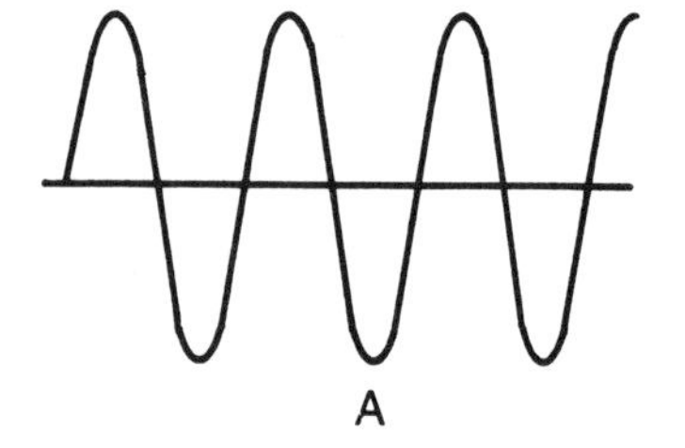

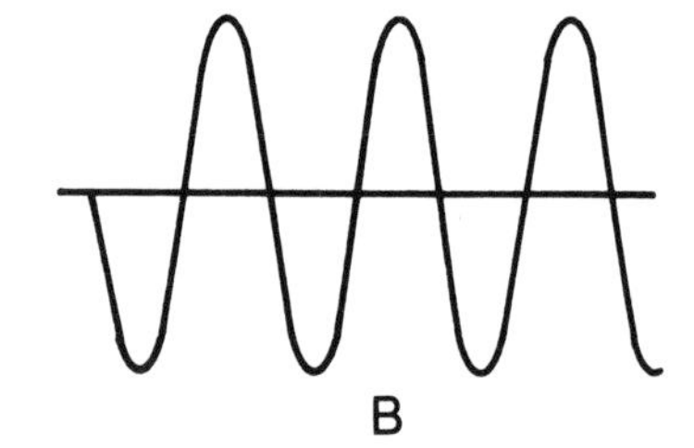

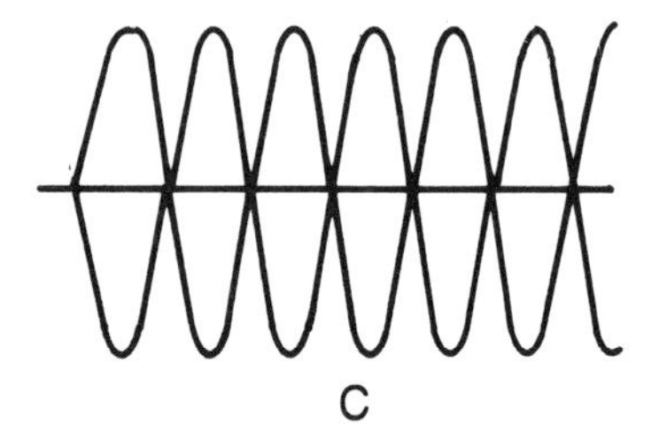

Fig. 11-4. When you combine a signal (*A*) with its inversion (*B*) the two cancel each other out and you get . . . nothing (*C*).

Common Sense

The word "common" appears frequently in electronic terminology. You come across terms such as "common ground," "common-emitter" (or "-collector," or "-base"), and even "common-mode rejection" all the time, or so it seems. If you know how the word is being used, these terms become more intelligible.

Originally, the word "common" meant "shared equally." In electronics, the word is used primarily in its original sense.

For example, a "common ground" is a ground system, perhaps the chassis of a piece of equipment, shared equally by all the components in a circuit. Occasionally a ground is just referred to simply as a "common," which leads to its appearance in a term such as "common-emitter." Note the hyphen. What it indicates is a connection, and not just between the two words. It indicates a connection between the emitter and the common ground.

Finally, in the term "common mode rejection" the word is used in its sense of "something shared." Usually, in this instance, what is shared is some form of unwanted signal, perhaps hum. Operational amplifiers have characteristics that make them particularly well suited for use in circuits that eliminate (or reject) this type of interference.

background noise. You could feed that signal to one input of an op amp circuit, and feed another signal containing just the background noise—from an unused frequency near the one you were tuned to—into the other. Inside the op amp the two noise signals would cancel each other out, leaving only the intelligence-carrying signal to appear cleanly at the output. This is called common-mode rejection, the word "common" here referring to the fact that the noise is common to both inputs.

A VOLTAGE FOLLOWER

Figure 11-5 illustrates the simplest form of op amp circuit, an amplifier with unity gain called a *voltage follower.* The op amp is the only component—there's nothing else, not even a resistor.

A signal is input to the non-inverting input of the op amp, and the entire output of the device is fed back to inverting input. Because no differential voltage is developed (which is one of the "ideal" op-amp characteristics under these circumstances), the input and output voltages

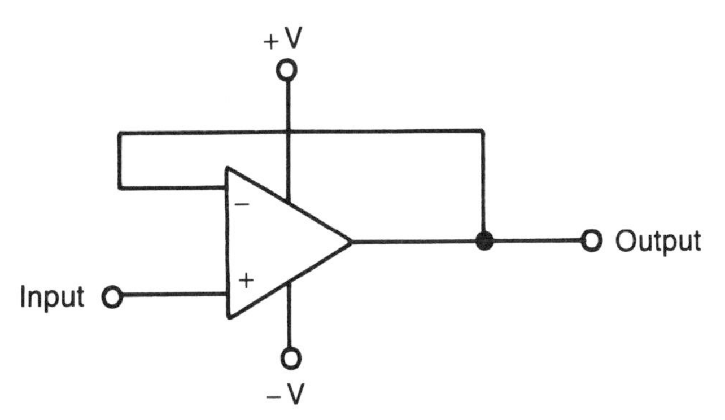

Fig. 11-5. This output of this voltage follower circuit is identical to its input, but can have more drive. The circuit can also provide impedance matching or isolation between other circuit stages.

are equal. This device is another version of the transistor unity-gain amplifier, or buffer, that we saw in Chapter 10. It can be used to provide buffering between stages, or sometimes for impedance matching.

OP-AMP AMPLIFIERS

A very simple amplifier built using a 741 op amp is shown in Fig. 11-6. It illustrates several practices that are followed in designing operational amplifier circuits. This is an inverting amplifier—that is, the positive portion of the input signal becomes negative at the output, and the negative portion becomes positive. The amplifier inverts the signal, or turns it upside-down. In many applications, it does not matter whether a signal is true or inverted. For reasons we'll see in a moment, this type of op-amp amplifier can be potentially much more stable than a non-inverting amplifier. If it is essential that the output of your amplifier be "true"—right-side-up, as it were—you can use two inverting stages to first invert the signal and then re-invert it so you are back where you started.

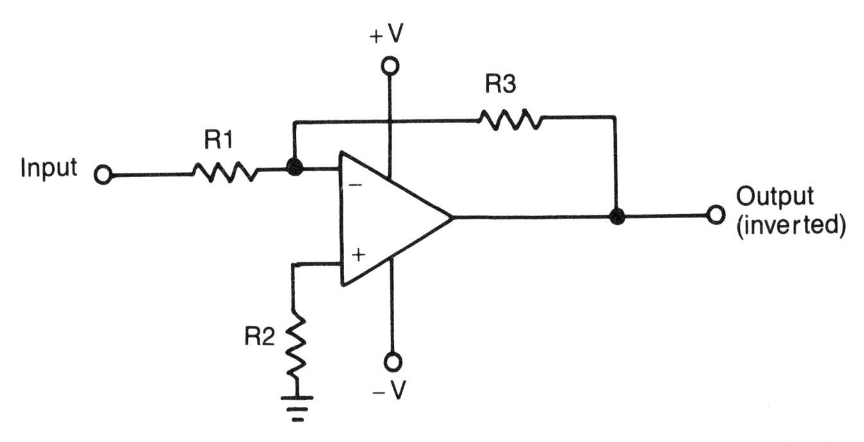

Fig. 11-6. The gain of this inverting amplifier is set by the feedback resistor R3. In similar inverting and non-inverting circuits, it is always this resistor that performs this function.

Resistors R1 and R2 are the input resistors. They are necessary because, like transistors (which is what op amps are made of) op amps are current-operated devices, and as is the case with all such devices (remember the LEDs in Chapter 9?) the input current has to be controlled. The value of the input resistors is not especially critical—values from 1k to 100k are commonly used, but it is a good idea to keep the resistances as low as is practical. In the inverting amplifier shown in Fig. 11-6, the value of R2 is equal to the combined value of R1 and R3 in parallel, which is calculated by:

$$\frac{1}{R3} = \frac{1}{R1} + \frac{1}{R2}$$

This causes the resistances in both input legs of the op amp to be equal. Resistor R3 is called the *feedback resistor* and is used to set the gain of the amplifier. By feeding back some of the output signal to the input, and since it has a polarity opposite that of the input signal, a portion of that input signal is canceled. The degree of reduction in input strength determines the degree of gain of the amplifier.

The amount of gain is equal to the ratio of R3 to R1, and the output of the amplifier is given by the equation:

$$V_{OUT} = V_{IN}\left(\frac{R3}{R1}\right)$$

The gain of the circuit in Fig. 11-6 is set to 100. Having somewhat arbitrarily selected a value of 4.7k for R1, that makes R3 equal to 470k, a hundred times R1. Substituting those values in the parallel-resistance equation we have:

$$\frac{1}{470{,}000} = \frac{1}{4700} + \frac{1}{R2}$$

since

$$\frac{1}{4700} = \frac{100}{470{,}000}$$

$$\frac{1}{R2} = \frac{1}{470{,}000} - \frac{100}{470{,}000}$$

$$\frac{1}{R2} = \frac{-99}{470{,}000}$$

In this case, we can ignore the minus sign, so:

$$R2 = 4747 \text{ ohms}$$

For all practical purposes, R2 turns out to be 4.7k, the same value as R1 and on many applications you'll find that the values of the two input resistors will be the same.

In extremely exacting situations you might want to balance the currents in the op amp's two legs precisely. For this purpose, some op amps (the 741 included) have two pins labeled OFFSET ADJUST. A resistor is intended to be connected between these pins to apply an offset voltage internally. Most of the time you can just ignore these pins—one of the few instances where you do not have to tie an unused pin high or low.

Plus and Minus

The universal symbol for an operational amplifier is the triangular one. If you see a triangular symbol in a schematic diagram, you're looking at an op amp. It may bear a fancier label—maybe "3-watt audio amplifier"—but underneath it's just a plain old op amp. Normally the devices's two inputs are shown at the left, and the output at the rightmost tip of the triangle.

The inputs are marked, one with a "+" and one with a "−." While you can think of the minus sign as standing for "inverting" and the plus sign as representing "non-inverting," their actual meanings are something else—"positive" and "negative," just as with a dry cell or other source of electricity. Operational amplifiers are current-operated devices, and there is a current flow between the two inputs. Current flows into the negative input and out of the positive one (even though it's still called an input).

The nature of op amps dictates that "what goes in must come out," so the currents flowing in both input legs of the device must be equal. If they are not, a differential situation arises. We can frequently make use of this differential (as is the case in using feedback resistor to control the gain of the device) but in general, the goal is to have equal resistances (and therefore equal currents) in both legs.

A non-inverting op amp is shown in Fig. 11-7, and it's not too much different from the inverting version except, of course, that its output is "true;" that is, the polarity of the output signal is the same as that of the input signal. It does not, however, use an input resistor. That resistance is assumed to be provided by the stage of the circuit preceding the op amp, and R1 in the inverting leg of the circuit should be equal to that resistance. The gain of the amplifier is set by resistor R2, the feedback resistor, and is equal to:

$$\frac{R1 + R2}{R1}$$

If R2 is large compared to R1, the gain will be approximately equal to the simple ratio of R2 : R1. The fact that the input resistance of the

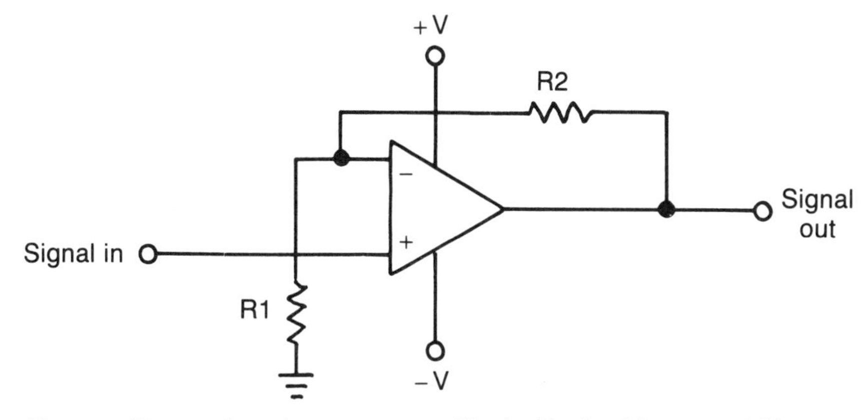

Fig. 11-7. The non-inverting op-amp amplifier in this circuit is not as stable as an inverting one because its input impedance is dependent on the circuit stage preceding it.

non-inverting leg is provided by a previous circuit stage, and may not be constant, can result in this type of amplifier being less stable than the inverting type.

COMPARATOR

Finally, in Fig. 11-8 is another extremely simple op-amp circuit known as a *comparator.* As its name implies, a comparator compares—in this case it compares the voltage at one of the op amp's inputs with the voltage at the other. If the voltage input through resistor R2 exceeds the reference voltage applied through resistor R1 to the other input of the IC, the output of the device will immediately swing to the level of the supply voltage.

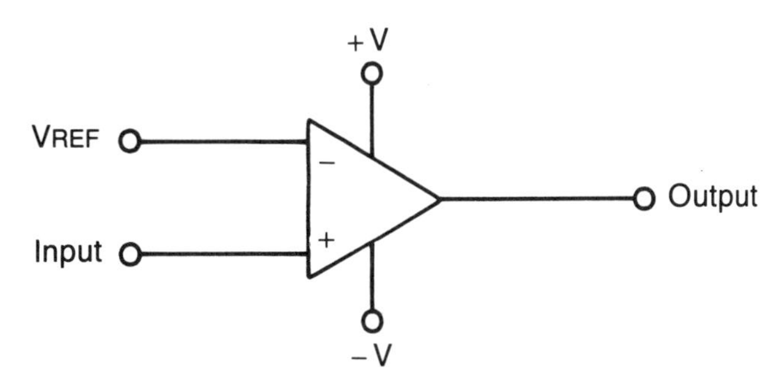

Fig. 11-8. The output of this comparator goes low when the input voltage equals the reference voltage. This makes it useful as the heart of a warning device (among other things).

This type of circuit can be used to give an indication of when a certain voltage level has been reached. For example, the output of the op amp could be used to illuminate an LED indicator. If the op amp is powered by TTL-compatible voltages (five volts) it can be used to drive TTL or other logic-family ICs.

Strings of comparators that are weighted with resistors (Fig. 11-9) can be used as analog-to-digital converters. The resistors in this circuit

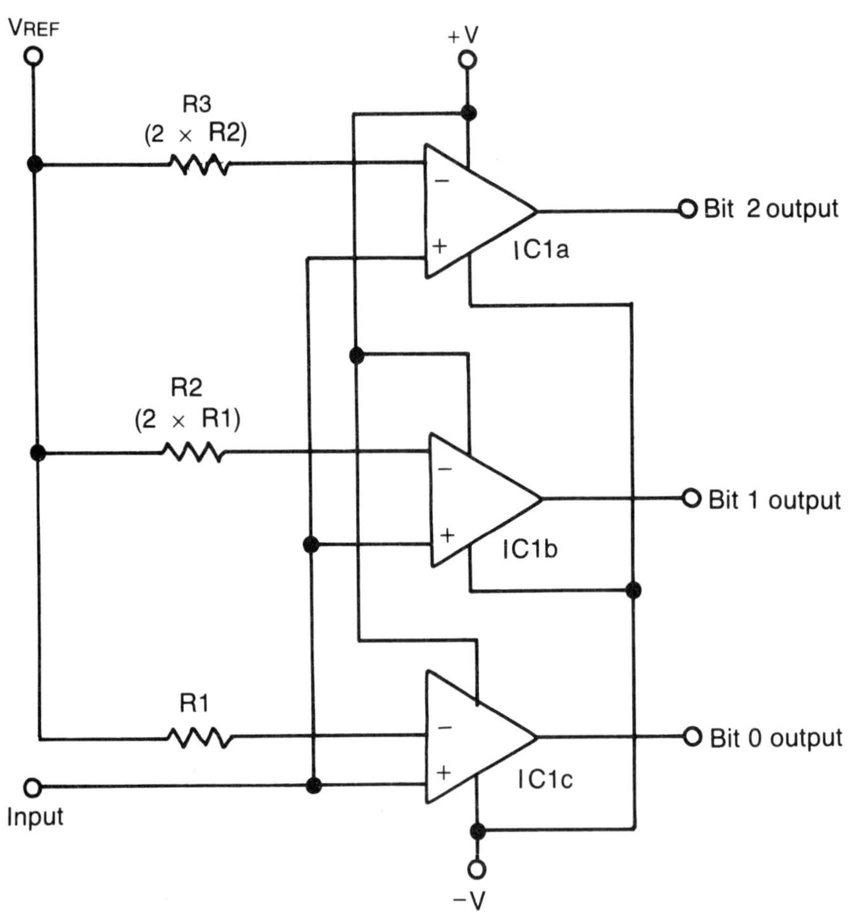

Fig. 11-9. Comparators can be connected so their outputs form a binary word that can be used by digital circuits. The same signal is input to all the devices, and a series of resistors, each with a value twice that of the preceding one, establishes the reference voltage of each stage at half that of the preceding stage. This arrangement of resistors is known as an R-2R ladder, referring to the way the values increase. The terms "IC1*a*" through "IC1*c*" refer to three of the op amps that are contained in a multi-device package (dual and quad op amp packages, for example, are quite common, and cost little more than single-device packages).

form what's known as an *R-2R ladder*—the resistance (and thus the voltage required to overcome it) associated with one stage is twice that of the previous one. The individual outputs, then, are weighted so that each represents a power of 2 (2^0, 2^1, 2^2, etc.) and the output as a whole takes the form of a binary number, where each bit represents a value twice that of the next lower one. The output of a circuit such as this can be put almost directly into digital memory or manipulated by digital logic circuits.

12
Electronic Switches

Several times in the course of this book the subject of bias has come up. Applying a voltage to a semiconductor junction to enable, or occasionally to inhibit, current flow is a basic principle of operation of most solid-state devices. At its simplest, applying a bias voltage to a junction can result in an electronic switch.

You can make a simple electronic switch with just a diode and a $1^1/_2$-volt dry cell or other low-voltage dc source. In the circuit illustrated in Fig. 12-1 a low-level audio signal, perhaps the output of a preamplifier, is applied to the cathode (−) end of a diode. While just about any diode will do the job, a 1N914 or 1N4148 switching diode is used here. The anode (+) end of the diode is connected to the input of an amplifier. That end of the diode is also connected, through a switch, to the positive (+) end of the dry cell.

As long as that switch is open and no forward bias (the $1^1/_2$ volts from the dry cell) appears across the diode no audio signal will pass. When the switch is closed, however, the diode will become forward biased and the 0.3-volt ''hurdle'' will be overcome. The diode will now conduct, and the audio signal, as you'll be able to hear, will be able to pass through the diode and onto the rest of the system.

Of course, with a mechanical switch available, you don't really need an electronic one. However, an electronic switch has the advantage that it can be controlled directly by other electronic devices, without the need for expensive (and eventually unreliable) mechanical devices. For example, the bias voltage could be provided by the output of a logic cir-

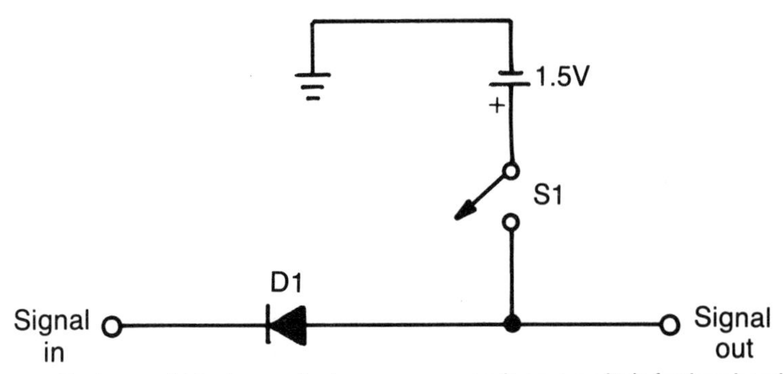

Fig. 12-1. By forward-biasing a diode, you can use it as a switch for low-level signals. If bias is present, the signal passes; if it is not, it's blocked.

cuit. As illustrated in Fig. 12-2, a signal from one output of a multi-output logic device could be used to activate a particular diode switch and allow an audio signal from a specific source to flow.

SCRS AND TRIACS

One step up from diodes, perhaps, on the ladder of sophistication, are the devices known as silicon controlled rectifiers, or SCRs. These are *latching* diode switches—once they are turned on and start to conduct, they will stay turned on even when the signal that was used to trigger them is removed. SCRs come in all sizes, from little ones with current-handling capabilities of a few hundred milliamperes to large industrial-size units that can handle currents of dozens of amps and, in the case of some units designed for use in radar transmitters, control short pulses on the order of 1000 amperes.

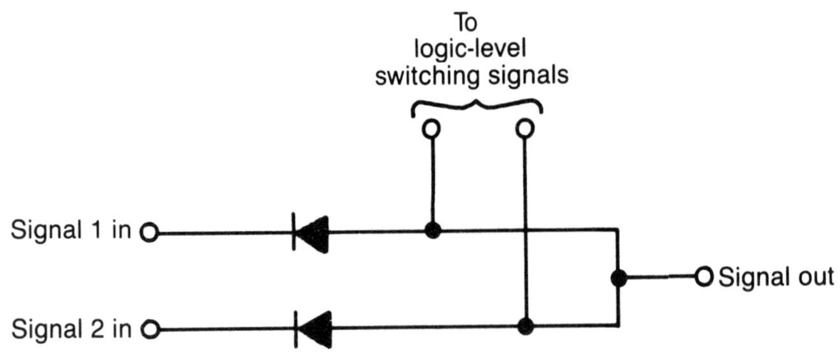

Fig. 12-2. Diode switching is easily accomplished using the outputs of logic-level devices.

The schematic symbol for an SCR is shown in Fig. 12-3. An SCR is connected as would be a diode and the trigger signal is applied to the third lead of the device, called the gate. There are two families of SCRs, each having a different gate sensitivity. The standard-gate type is intended to operate from a trigger current on the order of 40 mA or so. *Sensitive-gate* devices, which are the type you'll most commonly encounter, operate with trigger currents of much less than a milliampere—sometimes as low as 200 microamperes—0.2 mA!

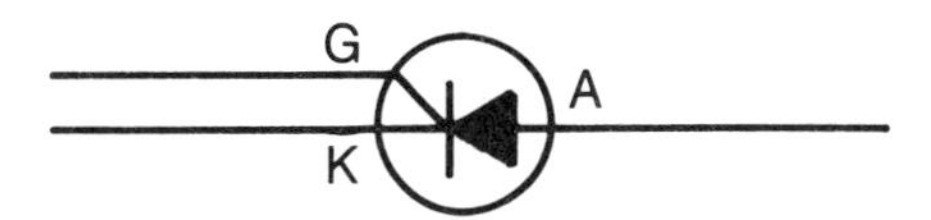

Fig. 12-3. A silicon-controlled rectifier, or SCR, conducts when a signal is applied to its gate (indicated by the letter "G"). It latches, and continues to conduct even when that signal is removed.

SCRs are rated, like ordinary diodes, according to their power-handling capacity in amperes and by voltage. For a given current-handling capacity there is usually a selection of several devices having different maximum working voltages.

Figure 12-4 shows a simple SCR circuit that allows you to turn on an LED with a momentary switch and then see it remain on even when the switch is released. The SCR will continue to conduct, and the LED remain on, until the circuit supplying power to the LED is broken, not to the gate; it needs only a short initial pulse to start the device conducting. Only a momentary break in power is required to reset the circuit. Because SCRs are current-operated devices, a current-limiting resistor is necessary in the gate circuit. For sensitive-gate devices a resistor with a value of about 27,000 ohms will work nicely.

Silicon controlled rectifiers are direct-current-only devices. If you were to try to control an alternating current with an SCR, it would act like an ordinary rectifier and block half the ac waveform. However, two SCRs connected back-to-back in parallel, as illustrated in Fig. 12-5, will conduct both halves of an ac waveform, each of the devices handling half the work. When a pair of back-to-back SCRs is put into a single package with a common gate, the result is a *Triac*. Triacs are used to control alternating current. They may be used for simple on-off control of an ac device or with a few additional components, for more elaborate functions such as motor-speed control or light dimming. The popular dimmer switches for room lighting use Triacs, not potentiometers as you might

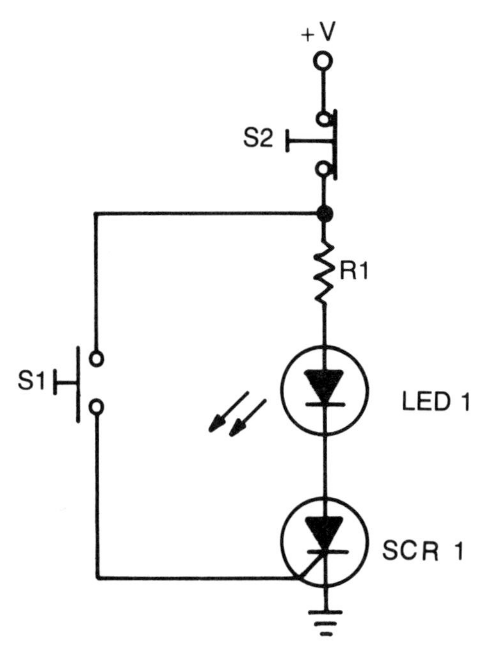

Fig. 12-4. The LED in this SCR circuit will light when switch S1 is depressed, and stay lit after it is released. Switch S2 is used to interrupt power flow to the circuit; when this is done the LED does—of course—turn off.

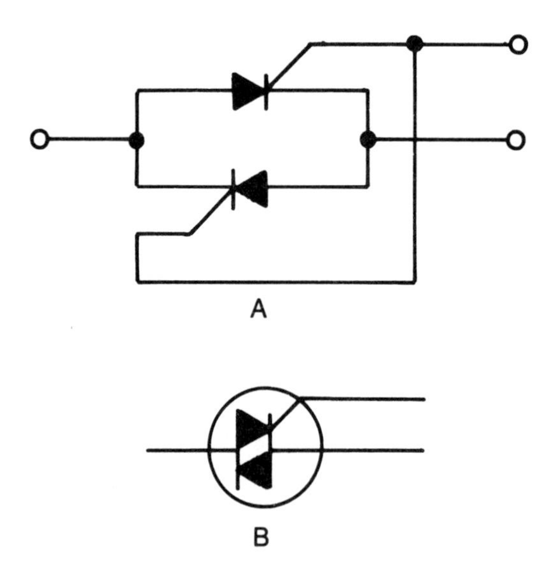

Fig. 12-5. A single SCR conducts only direct current. Two SCRs connected back-to-back (*A*) conduct ac. Such a two-SCR device is known as a Triac; its schematic representation is shown in *B*.

have thought. Triacs are much more efficient in that they do not dissipate unwanted electricity as heat; they simply block its flow.

LIGHT-DEPENDENT DEVICES

Light and electricity go hand-in-hand. Where there's one there's frequently the other. In devices such as LEDs, electrons give rise to photons. And, in devices such as photocells, light causes an electric current to flow. There's something about light-operated circuits—maybe the fact that you can actually see the cause and the effect—that makes them especially intriguing to build and tinker with.

Devices that use light to control the flow of electricity work in one of two ways: they either generate electricity when struck by light, or their resistance changes making them a kind of variable resistor. Cadmium sulfide (CdS) photocells, an example of one is shown in Fig. 12-6, are an example of the latter type. In the dark, the resistance of a CdS cell is high. When it is struck by light, however, its resistance decreases dramatically. The device generates no current; it is used strictly as a light-dependent resistor.

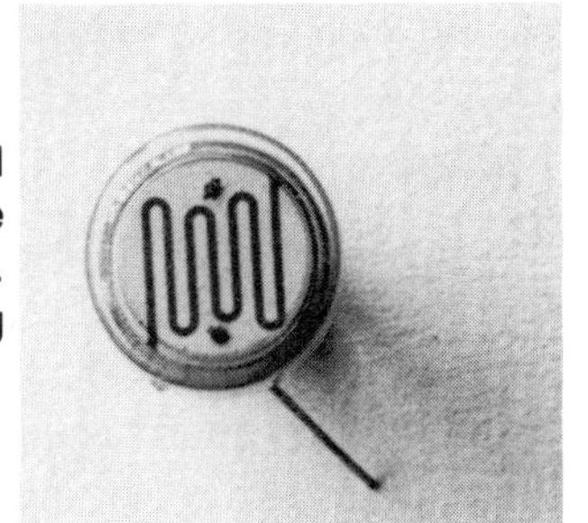

Fig. 12-6. A cadmium sulphide (CdS) photocell acts as a light-dependent resistor whose value changes with the intensity of the light falling on it. This type of device requires a current flowing through it to operate.

This type of light sensor requires a power source so that a voltage drop can be developed across it. As its resistance varies in response to the intensity of the light striking it, that voltage drop will also change. This changing voltage can be applied to the base of a transistor amplifier, whose output can be used to drive a meter (to make a light meter), or to reproduce the audio (or other) information carried by a modulated light beam.

The other class of photoelectric devices actually generates an electrical current when struck by light. Devices such as these can be simple silicon photocells, more versatile phototransistors, or even more elaborate devices such as optocouplers. All work on the basis that when they are struck by light they produce an electric current.

Many photoelectric devices work both ways. If a device generates light when a voltage is applied across it, it will generate a voltage when exposed to light. Try connecting a voltmeter across the leads of an ordinary LED (Fig. 12-7). When you expose the LED sunlight, it will develop a potential well in excess of one volt across it. Maybe you'll remember this sometime when you're really stuck for a light sensor!

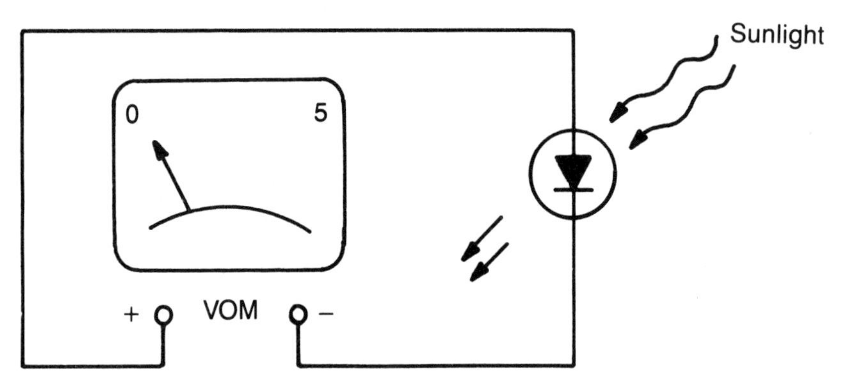

Fig. 12-7. When current is passed through an LED, light results. When light is applied to an LED, a voltage results.

Phototransistors

If you were to crack open a transistor and shine a light on it, you'd find that it, like an LED, develops a potential between the emitter and collector. (If you could do this carefully, you could still use the device; transistors are not vacuum tubes and when you break them open you don't ruin them by letting air in, you ruin them by destroying the very delicate connections to the semiconductor materials.) Fortunately, you don't have to demolish a dozen or two transistors to demonstrate this. A device called a ***phototransistor*** comes with a transparent plastic window to its inside, or sometimes encased entirely in clear plastic to let the light in.

Figure 12-8 shows the schematic symbol for a phototransistor. It looks pretty much like an ordinary garden-variety transistor except for the two arrows pointing to it; these arrows symbolize light striking the device (in contrast, the symbol for an LED has the arrows pointing away from it to indicate light being generated by the device). Phototransistors often have only two leads, one from the emitter and one from the collector. The base of the transistor is usually not connected to anything; the emitter-base forward bias is generated when light strikes the junction.

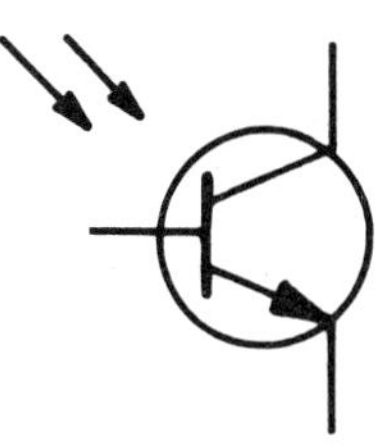

Fig. 12-8. The base lead of a phototransistor is frequently omitted—both schematically and in practice—since it is often not needed.

Working with Infrared Light

Most phototransistors are much more sensitive to infrared than they are to visible light. They are often sold with "matched" high-output infrared LEDs that generate light of approximately the same wavelength as that at which the response of the phototransistor peaks.

Working with infrared light presents a couple of problems. The first, of course, is that you can't see it. You have no direct way to tell whether your infrared LED is illuminating when you think it is. The solution to this is simple: have your output circuit drive *two* LEDs: one infrared, and another one that's red, green, or amber and that you can see. Such a circuit is shown in Fig. 12-9.

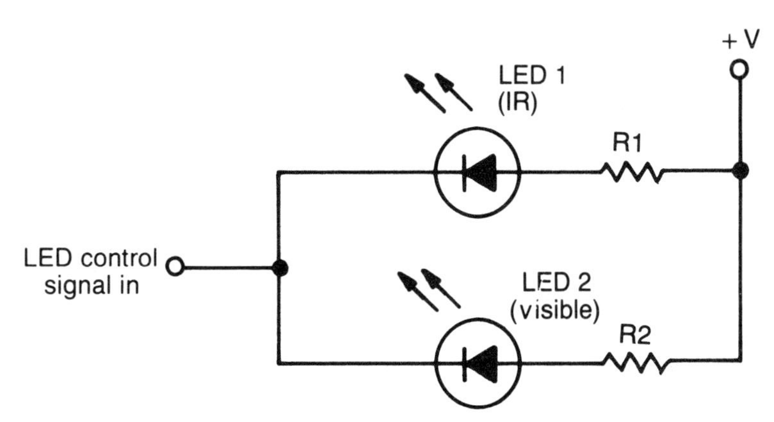

Fig. 12-9. To tell when an infrared LED is on, connect another, visible-light, LED in parallel with it to the same signal source. Each LED must have its own current-limiting resistor.

Note that some high-intensity infrared devices can draw two or three times more current than the 20 milliamperes or so to which you would normally limit an LED. Check the specifications on the device you use, and adjust the current-limiting resistor(s) accordingly.

Another source of difficulty arises from the fact that although infrared phototransistors are most sensitive to infrared light, they are also sensitive to other wavelengths (in the visible spectrum) as well. In other words, an infrared phototransistor can become "confused" by light coming from a source other than yours. There are two things you can do to prevent this. The first is to keep the light sensor as much in the dark as possible, possibly enclosing it in an open ended tube of some sort.

The other thing is to add an infrared filter to its light path. Filters of this type are used on wireless infrared remote controls and in front of the sensors upon which they are trained. You can purchase infrared filter material from optical supply houses, but it isn't cheap. Instead, try using the exposed and developed "tail" from a roll of Kodachrome film. It may look black, but it's transparent to infrared. Just make sure the film is Kodachrome; other color films, or black-and-white types, won't do the job. This filter will keep visible light from affecting the phototransistor. However, you will still have to take care to isolate it from other sources of infrared radiation such as direct sunlight.

For experimental purposes you can enclose both the infrared LED and phototransistor in the same light-tight tube, one device at either end facing inward (Fig. 12-10). Seal the ends with construction paper or some other opaque material, and bring the leads out through the paper or the side of the tube. Use clamp- or alligator-type clips to connect the devices to your transmitter circuits. This, by the way, is a homebrew version of a device called an optocoupler.

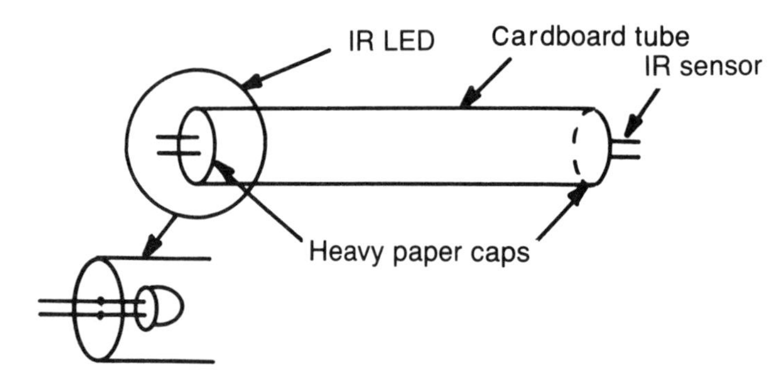

Fig. 12-10. To keep unwanted light (visible and infrared) from interfering with your IR sensor, enclose it and the IR source in a cardboard tube. This is a home-brew form of a device called an optocoupler.

Phototransistor Circuits

When light shines on a phototransistor it becomes forward biased and conducts. Phototransistors are not used for amplification, simply as receptors. Depending on how they are biased, phototransistors can be simple switches or can detect and pass on signals of varying strength — such as light modulated by sound—for amplification.

The phototransistor in the circuit in Fig. 12-11 is configured as a simple light-operated logic-level switch. When Q1, the phototransistor, is struck by light it becomes forward biased and permits the collector voltage, 5 volts, to flow to the emitter. This voltage present at the emitter can then be used to operate devices requiring logic-level inputs.

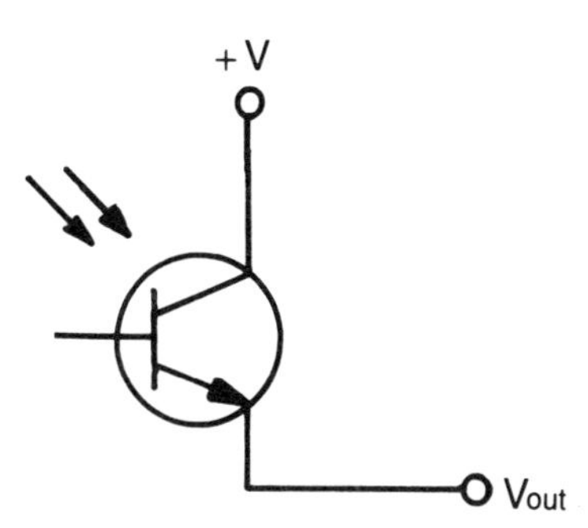

Fig. 12-11. A phototransistor can be used as a light-operated switch.

In Fig. 12-12, an infrared phototransistor is biased by resistors R1 and R2 so as to act as a sensor of light variations. The amount of current that flows from the emitter is proportional to the amount of light falling on the junction. Since sunlight is rich in infrared radiation (that's why you get hot standing out in the sun in summer) you can achieve some interesting results by connecting a sensing circuit such as this to an audio amplifier and placing the infrared sensor in a place where it will receive varying amounts of sunlight—in the shade of a leafy tree, for instance, where breezes will cause the amount of light falling on the sensor to vary.

LASCRs and Optocouplers

Light-Activated Silicon Controlled Rectifiers, or LASCRs (pronounced "LAS-kers") are switching devices but, like phototransistors, they are activated by light rather than directly by an electric current. The schematic symbol for a LASCR is the same as that used for an SCR except for the arrows used to indicate that it is a light-operated device.

LASCRs are not intended to switch large currents. Generally they are used to trigger other, heavier-duty, switching devices such as larger SCRs or Triacs. They are also found in optocouplers.

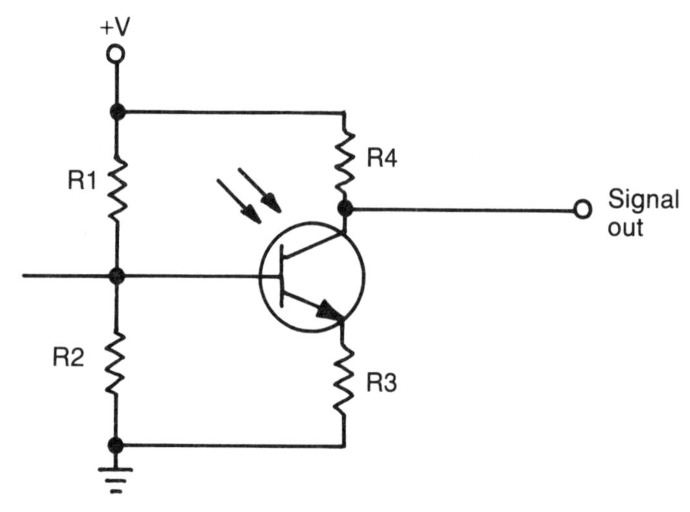

Fig. 12-12. By biasing a three-lead phototransistor the same way you would an ordinary transistor, you can construct a photosensor that fits right into the rest of an amplifier circuit.

An *optocoupler* is the all-in-one-package version of the phototransistor-LED arrangement in a single tube that was described above. Optocouplers are often called optoisolators because there is no electrical connection between the electronic input and output devices. The only connection is the light generated by the LED on the input side and received by a phototransistor at the output side of the device. This electrical isolation prevents signals or unwanted voltages from one end from getting into the circuitry at the other. As the schematic symbol for an optically coupled Triac switch in Fig. 12-13 shows, the package consists

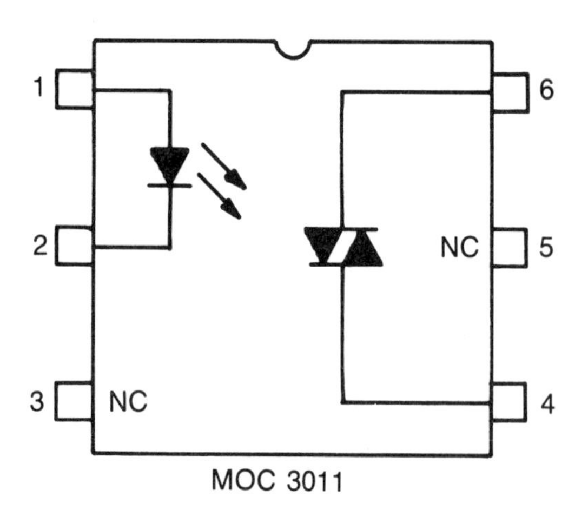

Fig. 12-13. An optocoupled Triac such as the MOC3011 can be used for switching where electrical isolation between stages is required. The light-sensitive Triac is made up of two LASCRs back-to-back.

of an LED and a light-activated Triac. A Triac, you'll recall, is nothing more than two SCRs (or LASCRs, in this case) back-to-back in parallel.

Optoisolators frequently come in the six-pin DIP package shown in Fig. 12-14. This type of six-pin package uses the same 0.1-inch lead spacing as other DIP packages, and can be inserted into an eight-pin socket with two pin-positions left over.

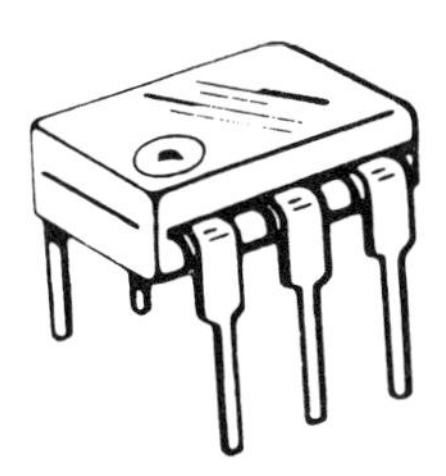

Fig. 12-14. Optocoupled Triacs come in six-pin DIP packages. Since six-pin DIP sockets are hard to locate, use an eight-pin one.

PART THREE

13
Troubleshooting

There comes a time when, finally, your design tools laid aside, and your soldering iron cooling, you turn on the fruit of your labors . . . and nothing happens! *Nothing*! No lights, no sounds, not even a shower of sparks. Now what do you do?

What you do is to start looking for whatever's keeping your project from coming to life, locate the problem, and correct it. And, while it's certainly disappointing that your design didn't work the first time, in finding out why it didn't work you may come up with a better design along the way. And, believe it or not, troubleshooting (if you put the initial disappointment and frustration aside and concentrate on the task at hand) can be challenging and interesting.

KISS

The first fact you have to recognize as you begin troubleshooting is that the problem you're trying to locate is probably not due to a defective component or to any fault attributable to anyone other than yourself. The reason your device doesn't work, or doesn't work correctly, is almost certainly your fault. While a defective part might prove to be the culprit, the cause of your difficulties is more probably something you did, or didn't, do. It's quite likely, too, that the problem will turn out to be a big and easily traceable one, and not one of those niggling little things you think is going to take a half a week to track down.

Start with the obvious. Or, what *should* be obvious. The acronym KISS stands for "Keep It Simple, Stupid." And, regardless of your IQ, that's what you have to do. Don't try to complicate an issue that's probably quite elementary.

For example, is the unit plugged in? Of course it is, you remember doing it. Check anyway. Maybe you did, and maybe you didn't—maybe you remember thinking about plugging it in. Or perhaps you really did plug it in, but accidentally unplugged it when you moved the device. Or maybe someone else unplugged it in your absence and forgot to tell you so. It certainly doesn't hurt to check. And something as obvious and easy to do as this certainly beats getting out all your test equipment, dismantling half the project (you didn't button it up inside its case before you checked it out, did you?) and trying to decide where to start. It is far more frequently the excruciatingly obvious that is the source of your troubles than the more obscure. It probably isn't even your design that's at fault.

Obvious Things

It is generally easiest to start at the outside of things and work your way inward. The unplugged plug is an example of this. (If the device is battery operated, did you remember to connect the batteries? Are they fresh?) If there's no output, is the output device—the speaker or LED, for example—connected?

Alright, let's assume that both ends of the device, input and output, seem to be in order. How about the fuse? To begin with, did you remember to insert the fuse in its holder? Assuming you did so, is it still intact? You can usually tell a blown fuse by its melted or internally scorched appearance. Sometimes, though, a fuse will look OK even though it has blown. The break in the element might have occurred where you can't see it. Check the fuse with your ohmmeter for continuity. If it proves to be open, first check its value to make certain that it was not too small for the current you intended to pass through it. Also, if the circuit is one that draws a larger current initially than it does after it settles down to normal operation (charging up a very large capacitor in a power supply would result in such a surge current) be certain that you used a slow-blow type fuse intended specifically to handle that initial surge without blowing.

Visual Checks

Before you even get out your ohmmeter, soldering iron, long-nose pliers and other instruments of torture and coercion, there is plenty you can do without them to search out the cause of your problem. Remember, whatever it is that's keeping your device from working is probably—

probably, mind you—not something that would require the resources of the Los Alamos National Laboratories to discover. Start your search with the naked eye.

Scrutinize your work. Examine it very carefully. There are three main problem areas you want to examine:

- Components
- Incorrect connections
- Poor soldering

Components and Connections

If, when you turned your device on for the first time, there was a flash of light and a sharp crack, followed by an acrid stink wafting up from it, something blew up. It may have been a resistor that became overloaded (although this is usually a problem that does not manifest itself immediately), an electrolytic capacitor that was installed backward, or an IC or other semiconductor device that was installed incorrectly, overloaded, or maybe was defective in the first place.

Sometimes, especially in a complex device where components are found on several circuit boards or are simply spread out, you can track down a difficulty such as one of these using your sense of smell to help you zero in on the problem area. The aroma will linger even after the time of the ''explosion'' is long past. And once you're in the neighborhood, you may be able to locate the problem area visually, even to the extent of pinpointing the component(s) involved.

Overloaded resistors look . . . well . . . burnt. They turn brown, as do the markings painted or inked on them, which makes them difficult or impossible to read. In extreme cases the body of the resistor cracks wide open. Don't be afraid to probe gently with the tip of some pointed object (such as a pencil or screwdriver) to make certain that nothing like this has happened. Do not, of course, do this with the power turned on, or even with the device plugged in if it is house-current operated. Until you have to begin actual electrical or electronic measurements, the device in question should be isolated from any electrical source that might cause further damage to it, or to your probing fingers.

Electrolytic capacitors that have exploded usually look exactly as though that's what they've done. They may bulge (the result of the sudden buildup of internal pressure) or they may have truly exploded and display a jagged hole or tear where the can ruptured. Most electrolytics manufactured today have a safety seal that's supposed to blow out if the internal pressure increases beyond a predetermined point, making such violent ends less common than they might otherwise be.

Integrated circuits and transistors can be damaged by too much voltage or current, and when they let go sometimes it is with a loud bang. And while violent explosions have been known to take place leaving no doubt as to the scene of the damage, it is often not possible to recognize a damaged semiconductor by visual inspection. It doesn't hurt to look, though. And once you smell what's left of an IC that's met a violent end, you'll never forget it!

Overheated components often leave their mark on circuit boards or on other components in their immediate vicinities in the form of scorching. Keep your eyes open for this symptom of trouble—both past and to come. A scorched area in a part of a circuit that is functioning indicates that it is not functioning *properly*. Something is overheating and will sooner or later lead to trouble. Check out these areas to find out what the problem is, and correct it before it gets worse.

Usually your problem will be less sensational and less obvious than one resulting in damage of a violent nature. In such a case, you have to put aside the dramatics and buckle down to a little less exciting investigative work.

The first thing to do is to make sure that there are no wiring errors, or PC-board design errors. With a copy of the schematic in hand (you did draw a schematic, didn't you? You weren't foolish enough to try to keep everything in your head?) start tracing all the connections from point to point, from component to component. Using a colored pencil for visibility, make a mark through each line that is connected correctly. This will keep you from checking the same place two or three times, something you may find yourself doing otherwise, particularly after you've been at it for an hour or two. If you have access to a photocopy machine that can make enlargements, make a blowup of the schematic. That will make things easier to see, and your markings and other notations won't tend to run into one another.

As you go from point to point checking the routing of control, supply, and signal lines, check too on the type and value of each component (and check each off on the schematic as you go) and on its orientation. Remember that if you are working on both sides of a circuit board, it is easy to become confused about which way the pin numbering on ICs and other multi-lead devices runs. It is not too difficult to confuse pin 14 with pin 1 and wire up an entire IC backwards. Even three-lead devices such as transistors can become confusing, particularly since not all case styles have leads in the same order.

Remember, too, that devices such as diodes and electrolytic capacitors, including the dipped tantalum types, are polarized and have to be connected in a particular way. Electrolytics, for example, have their shells (the "–" side) connected to the part of the circuit with the lower potential, which frequently means to ground. Similarly, the banded cathode end of diodes is usually the lower-potential end. Make sure that both the actual circuit, and the schematic representing it, are correct. Errors frequently creep into schematics, even the ones that are published in magazines and books, and you would do well to verify the circuit diagram yourself. If you are working from a schematic that appeared in a back issue of a magazine it's a good idea to check the issues published for several months after the appearance of the one containing the schematic to look for letters pointing out errors in, and corrections to, the circuit.

Semiconductors of the same type, particularly integrated circuits in DIP packages, tend to look alike. Make certain that the correct part is in the correct location.

And, of course, double check your readings of the color codes on such things as resistors. When you examine them, do so under a bright light so you can differentiate colors that can otherwise be confused with one another. For example, if your resistors came from several different sources, as they probably did, you may discover that one manufacturer's red is another company's orange. It's been known to happen. If you find such a questionable component, you may want to lift one end of it from the circuit so you can test it with a ohmmeter for the correct value. (You can't measure the resistance in-circuit—you would actually be measuring the combined value of the resistor and the resistance of a lot of the rest of the circuit in parallel with it!)

SOLDER FAULTS

Poor solder joints are also to blame for many non-functional circuits. This is less the case these days than it was back in the heyday of vacuum tubes, where a total of six connections to a single pin of a tube socket was not uncommon; and where it was very easy to not get good solder flow to one or two of the bottommost of those. Even so, it is not difficult today, especially if you are new to the game or an old-timer in a hurry, to miss a connection or two, or to work so fast that you do not do a thorough job on each joint.

Check each joint to ascertain that (a) it was soldered and that (b) it was soldered correctly. With regard to the first it's surprising how frequently, especially if you're soldering rows and rows of IC-socket pins at a time, you can miss one or two pins while being absolutely certain that you got them all. People who wire-wrap their circuits usually check off each connection on the schematic as they make it (the way you checked them off as you made your after-the-fact inspection) to keep things straight and to make tedious troubleshooting less necessary. You probably won't follow this procedure as you solder, but you might wish you had. If you start finding more than a few such careless omissions, try to work more slowly in the future. The time you invest in building the circuit correctly from the start will certainly be less than that you would have to put into troubleshooting later on.

Cold solder joints—those grainy, frosty-looking, frequently blobby joints that can prevent a good electrical connection—are also the result of too much haste. A cold solder joint can be hollow inside and not contact all the portions of the joint you intend it to. The best way to avoid cold solder joints in the first place is to apply enough heat to the area to permit the solder to flow freely and wet all the parts of the joint, and to keep applying heat until the solder has reached and adhered to all the places you want it to.

If you come across a cold solder joint in your investigation, you can correct it easily by reheating it with the soldering iron. If the joint is blobby you might want to remove some of the excess solder. You can use desoldering braid or a solder sucker for this, or sometimes the excess will migrate to the tip of the iron, from which you can (and should) remove it promptly. If there is not enough solder to flow over the joint, add a bit more. Make sure the joint is smooth and a little shiny after it cools to solidity.

While it's best to avoid them, you might on rare occasions have one of those six-wires-to-one-point connections, particularly in power-supply wiring on a chassis where you use terminal strips. Wiggle the wires by hand or with long-nose pliers to make certain that they *don't* wiggle where they're supposed to be soldered. If there's any cause to doubt the solidity of the connection, reheat it and add a bit more solder if it seems required.

Sloppy soldering can easily lead to short circuits. These shorts may not be in areas where they will provide an immediate indication of their existence such as violent explosions or loud bangs emanating from a power supply, but might instead simply conduct a signal or control voltage to ground or another point other than it was intended, resulting in a non-response. These soldering faults fall into two categories: those caused by the solder itself, and those caused by other materials such as wire stands and resistor leads.

As you work, solder that does not go into making a joint accumulates on the tip of the iron little by little. This is one reason you should wipe the tip routinely on a piece of damp sponge or paper towel—the tip should be kept clean. Sometimes, though, such an accumulation clears itself simply by dropping (or being flung) from the tip, usually when you're not aware of it.

These solder splashes or splatters can result in shorts on your printed circuit boards. Commercially made boards are usually coated with a solder resist, a layer of material to which solder won't adhere, and that acts as an insulator. There are holes in the resist layer at solder pads and other places where solder is to be applied. If you make your own circuit boards, however, it is unlikely that they will have such a coating. A solder splash that bridges two traces on the board (hence the term "solder bridge") will short them out. And, while surface tension helps keep solder from spreading irresponsibly, it is quite easy to form solder bridges unintentionally between two adjacent pins of an IC socket.

Some of these solder bridges can be so small that they are nearly invisible. When you search, do so under bright illumination and have a magnifying glass handy.

Solder bridges extending from a joint can be removed by reheating them with the tip of a soldering iron and using the tip to draw the excess solder, if there's not too much of it, back into the joint. This will be particularly easy on resist-coated boards, although the technique works on non-resist boards as well. You might find it necessary, particularly in the case of IC-socket pins, to use desoldering braid to remove excess solder. You should be able to do this pretty rapidly; too much heat can cause the delicate PC board traces to lift from their substrate.

The other soldering-related cause of short circuits is pieces of wire where they aren't supposed to be. If you are using stranded wire take care to see that the strands are twisted together at the ends of each piece and that there are no stray "hairs" fanning out to touch other points. These should be twisted or bent back to form part of the joint they originate from (you should do this before soldering; afterwards the wires will be stiff and it will be nearly impossible to move them). If something looks suspicious, get out a magnifying glass and check it closely.

When you cut off the excess lengths of resistor or capacitor leads, or hookup wire, make certain that you get them all out. Hold onto them as you clip them; this will prevent them from flipping halfway across the circuit board, into your eye, or into a dark corner of the chassis where they will lie until you close it up (then they'll sneak out and short out the power supply). It's not necessary to account for each lead with the accuracy of a surgeon counting his sponges, but you should be sure that none of the clippings got away from you.

Before you consider yourself done with the soldering portion of a job (or with the troubleshooting part if that becomes necessary) inspect both sides of the PC board for loose pieces of excess wire, snippets that may have gotten soldered across points where they fell (don't laugh, it could happen to you late some night), and pieces that should have been clipped but weren't. These last can be an embarrassing source of short circuits when they come into contact with a part of the chassis after you put the cover on your project.

Finally, give the unit a good shake. It's easy to lose a lockwasher or nut when you're working on the hardware phase of your project. Any such dropped parts should be retrieved and accounted for before you continue; these little odds and ends have been responsible for numerous short circuits over the years. Unless you specifically designed your project to rattle when shaken, if it does you probably have something loose inside that shouldn't be. Find it.

TEST EQUIPMENT

If a thorough visual inspection does not show up a problem, or if it does but the device still doesn't function properly after you've taken care of it, then it will necessary to probe a little deeper. For this you will need one or two pieces of test equipment—a multimeter (digital or analog) and, if you can beg, borrow, or steal one, an oscilloscope. Since there are plenty of books that deal directly with using these devices, we'll refer you to them and concentrate on telling you what to look for with them.

The most basic piece of electronic test equipment is the multimeter. It combines three functions: the measurement of voltage, current, and resistance. When someone refers to a volt-, am-, or ohmmeter he generally means a multimeter. Analog multimeters (sometimes referred to as VOMs, for "Volt-Ohm-Meters," or in the case of antiques as VTVMs for Vacuum Tube Voltmeters") have a conventional meter with a swinging needle on their face. There are a number of scales against which the needle can be read, depending on what you are measuring and how much of it is involved. The better meters have a mirrored scale with a small reflective band that runs behind the needle. When you read the meter you read it so the needle blocks out its reflection in the mirror; this gives the most accurate reading of the scale by eliminating parallax.

Because of their mechanical nature, analog meters can be less accurate than their all-electronic counterparts (which we'll get to in a moment), they are particularly well-suited to making continuity checks—that is, testing for zero resistance, which indicates a connection or short circuit. It is simple to keep an eye open for the needle movement from one end of the scale (infinity, or an open circuit) to the zero-resistance end. Unless you are specifically looking for a very low, but definite, resistance value, you don't even have to read the meter to know what you have. And, the action is nearly instantaneous.

Digital multimeters, also called DMMs (and sometimes DVMs, for "Digital Voltmeters" even though they also measure other quantities), have no moving parts. They convert their measurements into a signal that ultimately is displayed as a number on a liquid-crystal, or sometimes on a light-emitting diode, display. The value is displayed as a series of numerical digits, one, two, three, or four to the left of a decimal point, and one or two to its right. This type of display is extremely accurate—there's no parallax or estimation involved, all you have to do is read off the numbers—but it does have its limitations.

First of all, it is accurate only to the number of digits it can display. That is, if you're looking for four-digit accuracy (say, 3.141 volts) but the meter can show only one digit to the right of the decimal point, you have no idea of what the other three are. The meter will read only "3.1." This won't help you very much if you're looking for better accuracy than that.

The second problem with digital multimeters comes in using them for continuity testing. It takes time—not much, but still an appreciable amount—for the meter to count up or down to its final value and to display it. This time, combined with that fact that there is no swinging needle to catch your attention and give a quick indication, makes it easy to miss an indication of continuity if you're in too much of a hurry.

Most DMMs have a "continuity alarm" that sounds a tone when a resistance of zero ohms is encountered. This is convenient in that it frees you from having to keep an eye on the meter. However, it takes time for the meter to sound the tone, and by the time it's ready to do so you may already have moved on with your probes to test another spot, assuming that the meter didn't sound because it found some resistance there, not because you left before it could tell you there wasn't any. If you have to use a digital meter with a continuity function work slowly enough for the meter to have time to react.

Your budget permitting, you might want to invest in an inexpensive analog meter for YES-NO continuity testing, and digital one for determining voltage, resistance, and current values.

You might come across a device billed as a ''continuity tester'' in a hardware store, and even in some places selling electronic supplies. This is usually nothing more than a neon bulb with a couple of leads attached. These devices are intended to be used by electricians in working with house wiring; they have no place in your collection of tools.

Oscilloscopes

The other piece of test equipment you will eventually find useful is an oscilloscope, although it's not as essential in a beginner's toolkit as is a multimeter. While even the least expensive 'scope costs more than a good multimeter (although you may find a good buy in an older one at a ham or computer flea market), if you can justify its price it is a valuable tool to have.

While a multimeter deals with measurements of static or slowly changing values, a scope can handle and display rapidly changing ones. Except for measuring steady voltages or currents, multimeters are useless in ac circuits. They usually cannot tell you whether a signal is present, and can never show you what the waveform looks like.

Not only can you use a scope to check on signal waveforms, you can also use it to measure voltages. This makes it possible for you to see exactly what happens when you bias a circuit a certain way. Sometimes, for one reason or another, the wrong bias creeps in and blocks a signal from traveling through a device or circuit, or so alters it that is unusable.

When you use a scope to find out what's wrong with a circuit, don't just poke around at random. Work systematically from one end of the signal path to the other. Start with the input signal and follow it stage-by-stage through the circuit. At some point it will change drastically, or perhaps disappear. Somewhere in the area where this happens is your problem. Without an oscilloscope this type of signal tracing is extremely difficult, if not impossible.

Finally, an oscilloscope is invaluable as a learning tool as well as one for troubleshooting. By using its screen to observe what happens to a signal as you vary the values of the components in the circuit through which it passes, you can actually see what effects those changes have. There is no better way to learn.

Index